Advanced Structured Materials

Volume 135

Series Editors

Andreas Öchsner, Faculty of Mechanical Engineering, Esslingen University of Applied Sciences, Esslingen, Germany

Lucas F. M. da Silva, Department of Mechanical Engineering, Faculty of Engineering, University of Porto, Porto, Portugal

Holm Altenbach⬤, Faculty of Mechanical Engineering, Otto von Guericke University Magdeburg, Magdeburg, Sachsen-Anhalt, Germany

Common engineering materials reach in many applications their limits and new developments are required to fulfil increasing demands on engineering materials. The performance of materials can be increased by combining different materials to achieve better properties than a single constituent or by shaping the material or constituents in a specific structure. The interaction between material and structure may arise on different length scales, such as micro-, meso- or macroscale, and offers possible applications in quite diverse fields.

This book series addresses the fundamental relationship between materials and their structure on the overall properties (e.g. mechanical, thermal, chemical or magnetic etc.) and applications.

The topics of *Advanced Structured Materials* include but are not limited to

- classical fibre-reinforced composites (e.g. glass, carbon or Aramid reinforced plastics)
- metal matrix composites (MMCs)
- micro porous composites
- micro channel materials
- multilayered materials
- cellular materials (e.g., metallic or polymer foams, sponges, hollow sphere structures)
- porous materials
- truss structures
- nanocomposite materials
- biomaterials
- nanoporous metals
- concrete
- coated materials
- smart materials

Advanced Structured Materials is indexed in Google Scholar and Scopus.

More information about this series at http://www.springer.com/series/8611

Farzad Hejazi · Tan Kar Chun

Conceptual Theories in Structural Dynamics

 Springer

Farzad Hejazi
Department of Civil Engineering
University Putra Malaysia
Serdang, Selangor, Malaysia

Tan Kar Chun
Department of Civil Engineering
University Putra Malaysia
Serdang, Selangor, Malaysia

ISSN 1869-8433   ISSN 1869-8441   (electronic)
Advanced Structured Materials
ISBN 978-981-15-5442-1   ISBN 978-981-15-5440-7   (eBook)
https://doi.org/10.1007/978-981-15-5440-7

This Springer imprint is published by the registered company Springer Nature Singapore Pte Ltd.
The registered company address is: 152 Beach Road, #21-01/04 Gateway East, Singapore 189721,
Singapore

Contents

List of Figures

List of Tables

Chapter 1
Introduction

1.1 Static and Dynamic Equilibrium

Newton's third law of motion states that when a force is acting as an action, another force with equal magnitude will act in the opposite direction as reaction. A body is said to be in equilibrium when the net force acting on it is zero.

When that body is at rest, it is said to be in static equilibrium. An at rest body experiences forces from multiple directions, but since the net force is zero, the body remains at rest as described by Newton's first law of motion. For example, a stationary pendulum experience gravitational force and string tension force at the same time. However, the forces are equal in magnitude and opposite in direction and they cancel out each other, resulting in zero net force and the pendulum stays stationary (Fig. 1.1).

A body can be in equilibrium even when it is in motion, provided it moves in uniform velocity. The body is said to be in dynamic equilibrium under this condition. Let's say the pendulum drops because the string snaps. There are two forces acting on it throughout the fall: gravitational force and drag force, also known as air resistance.

Gravitational force is the function of mass and gravitational acceleration based on Newton's second law of motion, while drag force is the function of velocity. Since gravitational acceleration is constant at 9.8 m/s^2, gravitational force remains constant over the fall period.

The pendulum is not in dynamic equilibrium at first. At the moment of string snaps, the velocity of the pendulum increases from 0 to v_1. The drag force increases from 0 to $F_d(v_1)$ as well. However, the drag force magnitude is not enough to counter the gravitational force and therefore, downward acting net force pushes the pendulum, and its falling velocity increases. As falling velocity increases, the drag force increases, and downward acting net force reduces. This phenomenon continues until

© The Editor(s) (if applicable) and The Author(s), under exclusive license
to Springer Nature Singapore Pte Ltd. 2020
F. Hejazi and T. K. Chun, *Conceptual Theories in Structural Dynamics*,
Advanced Structured Materials 135,
https://doi.org/10.1007/978-981-15-5440-7_1

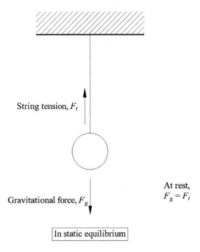

String tension, F_t

Gravitational force, F_g

At rest,
$F_g = F_t$

In static equilibrium

Fig. 1.1 An at rest pendulum

the magnitude of the drag force is equal to the magnitude of the gravitational force. At this moment, the falling velocity does not change anymore since the net force acting on it is zero. The pendulum is said to be travelling in terminal velocity, and it is in dynamic equilibrium (Fig. 1.2).

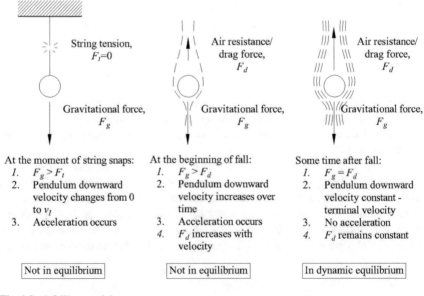

String tension, $F_t = 0$

Gravitational force, F_g

Air resistance/ drag force, F_d

Gravitational force, F_g

Air resistance/ drag force, F_d

Gravitational force, F_g

At the moment of string snaps:
1. $F_g > F_t$
2. Pendulum downward velocity changes from 0 to v_t
3. Acceleration occurs

At the beginning of fall:
1. $F_g > F_d$
2. Pendulum downward velocity increases over time
3. Acceleration occurs
4. F_d increases with velocity

Some time after fall:
1. $F_g = F_d$
2. Pendulum downward velocity constant - terminal velocity
3. No acceleration
4. F_d remains constant

Not in equilibrium

Not in equilibrium

In dynamic equilibrium

Fig. 1.2 A falling pendulum

1.2   Structural Dynamics

Structural dynamics is a study that evaluates and predicts structural response to dynamic load in the form of motion. In structural engineering, loads can be divided into two categories: static and dynamic. Static load is the load that acts in constant direction with constant magnitude at one point regardless of the time. Dynamic load on the other hand is the load that varies from time to time, either in term of direction or magnitude. Naturally, all loads on structure, including the self-weight of the structural member are dynamic load. This is because they were not applied at some point in time (Fig. 1.3).

However, not all 'dynamic' loads are able to excite motion. To do so, a load must be applied with high acceleration to overcome the inertia of massive structure and makes it vibrate. For this reason, an increase in dead load over the construction period is too slow to excite structural movement. Notable types of dynamic loads are wind, earthquake, impact, tide, machine vibration and vehicle movement.

Wind load is induced through rapid movement of air. When wind comes across a building, pressure will be developed on its windward face. The shape of the building often defines the effect of wind on the building. Earthquake load is induced through the movement of tectonic plates. Impact load is a high force or shock applied over a short time period when two or more bodies collide. Tidal load is induced through the movement of the oceanic water body. Analysis for tidal load is essential for marine structures such as jetty.

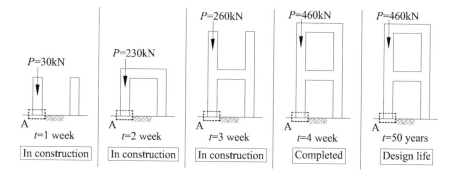

Say, P=30kN for column self-weight, P=200kN for 1-floor load transferred to column.

Total axial force acting on 'A', P_{ult}=460kN over 4 weeks.

Average increase in loading per day is merely **16.5kN/day**.

Fig. 1.3 Change of dead load with respect to time

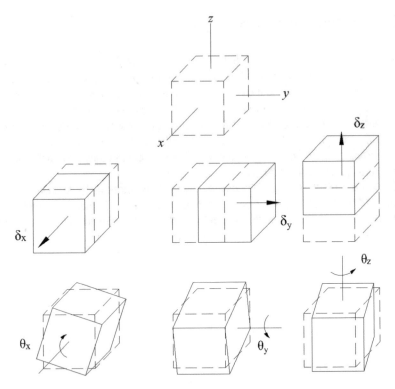

Fig. 1.4 Degree of freedom of a body

1.3   Degree of Freedom (DOF)

Degree of freedom is the possible component of transformation of a body. Funda-
mentally, the components of such transformation are expressed with respect to three
mutually orthogonal axes, namely x, y and z. For every axis, there are two possible
transformations: translation along that axis and rotation about that axis (Fig. 1.4).

1.4   Simple Harmonic Motion

Consider a spring–mass system: wooden block is connected to one end of spring,
and the other end of the spring is embedded in rigid support. When the wooden
block is displaced, the already extended spring pulls the wooden block up toward the
equilibrium position. When the wooden block reaches an equilibrium position, the
motion does not stop as the momentum associated with the motion is still exist. The
wooden block continues to go up until the momentum is diminished. Then, dominant

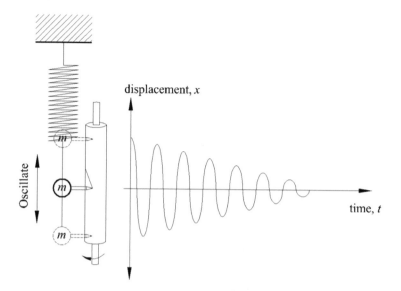

Fig. 1.5 Spring–mass system in simple harmonic motion

force acting on the wooden block is now a gravitational force, which is then being pulled toward an equilibrium position below it.

Simple harmonic motion is a form of periodic motion. If there is no energy loss due to damping, then the system will oscillate perpetually because the energy is conserved and converted to different forms at different positions. By recording the displacement of a wooden block with respect to time, we can produce a sinusoidal curve out of its smooth motion (Fig. 1.5).

Generally, the sinusoidal function of oscillator displacement, x with respect to time can be expressed as

$$x = X \sin \omega t$$

where X is the magnitude for displacement, t is time in second, s and ω is a frequency in rad/s.

Based on Euler's equation, sinusoidal motion can also be expressed in the following complex exponential form:

$$X e^{i\omega t} = X \cos \omega t + i X \sin \omega t$$

where e is $2.7\overline{1828}$ and i is $\sqrt{-1}$.

When the wooden block goes forth and back to its starting position (not necessarily the equilibrium position), we can observe that the corresponding sinusoidal curve completes a cycle. Since the oscillation is expressed in sine function, every complete oscillation shall equal displacement of 2π radians. Let's say the system takes time t to complete one oscillation, then ω is expressed as

$$\omega = \frac{2\pi}{t}$$

The unit for frequency in the above expression is rad/s. To express the frequency as number of oscillations per second, divide the above one with 2π rad/cycle:

$$f = \frac{\omega}{2\pi}$$

Period, T, defined as the number of oscillations completed in one second is equal to t since it is obtained by observing the time taken for the system to complete an oscillation. The correlation between T and f is as follows:

$$T = \frac{1}{f} = \frac{2\pi}{\omega}$$

Chapter 2
Components of Structural Dynamics

Under equilibrium, a structure subjected to dynamic load immediately develops reaction force from three components: mass, damping and stiffness.

2.1 Mass Component

Mass is defined as the body's resistance to acceleration. This characteristic of the body has been described by Newton's first law of motion: if the net force exerted on a body is zero, then the velocity of the body is constant, and its direction remains the same. In other words, the body is reluctant to change its travelling velocity, be it zero or a certain value, unless external load exerted on it.

Both Newton's first and second laws of motion are developed by ideally expressing a body as a point mass, where the exerted force will only cause linear movement, and deformation due to the force is zero. To apply the laws of motion, a body must exhibit these characteristics. Therefore, it must be assumed as a rigid body, where the force exerted is transmitted throughout the body and result in only linear movement. Distance between any points on the rigid body remains constant regardless of the force, differential translation at any points on the body does not occur and thus, no deformations take place (Fig. 2.1).

In real-life, structures do not exhibit ideal rigid body and point mass behaviour. First, point mass barely exists because, in a structure, mass is location-dependent. Take a reinforced concrete frame as an example. Reinforce concrete is heterogeneous, which means the mass at any point on it is different due to the random arrangement of its constituent materials: steel, aggregate and cement (Fig. 2.2).

A body does not really display rigid body behaviour when it is subjected to force either. At microscopic level, the presence of void between the body's molecules provides room for molecular movement, and deformation can be observed at the force exertion point, even for the most apparently rigid body.

© The Editor(s) (if applicable) and The Author(s), under exclusive license to Springer Nature Singapore Pte Ltd. 2020
F. Hejazi and T. K. Chun, *Conceptual Theories in Structural Dynamics*, Advanced Structured Materials 135, https://doi.org/10.1007/978-981-15-5440-7_2

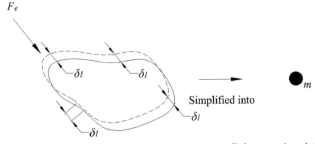

A high rigidity body with mass m experiences constant translation, δ_l over the body upon applied force F_e

Point mass (a point with mass m) permits only translation of body, and differential translation at between any two points is zero.

Fig. 2.1 Concept of point mass

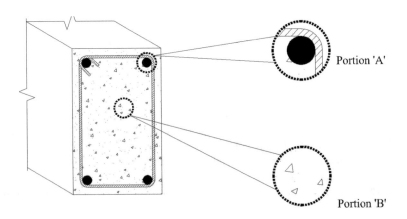

Mass, $M_A > M_B$ due to the presence of steel in portion 'A'

Fig. 2.2 Reinforce concrete as a heterogeneous material

For engineering purpose, however, variation of material and resultant behaviour at microscopic level are neglected because they are too small to bring significant effect to a building. In this sense, all materials are considered isotropic, assumes the body responds the same regardless of the location of force exertion. Ideal rigid body behaviour is possible under this consideration as well.

Therefore, a structure can be simplified. Take a reinforced concrete frame as an example. When lateral load is applied at the joint of structure, the structure sways. However, it is notable that slabs have very high rigidity compared to vertical structural element, i.e. columns, and they do not deform as much as columns. Therefore, slabs can be assumed as rigid diaphragm. By combining the dead load carried by the horizontal structural elements as well as the weight of the structure itself, every storey is considered as one rigid body and it can be expressed as a lumped mass, an idealized point that carries all the masses in one storey.

When subjected to force, lumped mass vibrates. Work is done and by work-energy theorem, kinetic energy increases by the magnitude of work done. Lumped mass responds to the force with acceleration since its velocity change from zero to a certain value after force exerts on it. Force is generated as a reaction in response to the applied force. The force is known as inertial force, F_I and it is expressed as the follows based on Newton's second law of motion:

$$F_I = \frac{\partial p}{\partial t}$$

Momentum is the product of mass and velocity. By substituting $p = mv$,

$$F_I = \frac{\partial (mv)}{\partial t}$$

Since mass is the amount of matter in a body, it remains constant over time. Thus,

$$F_I = m \frac{\partial v}{\partial t}$$

The rate of change of velocity over time is defined as acceleration, a. Thus,

$$F_I = ma$$

Meanwhile, velocity is also defined as the rate of change of displacement, x with respect to time. Thus, inertial force can be expressed in the following form as well:

$$F_I = m \frac{\partial}{\partial t} \left(\frac{\partial x}{\partial t} \right)$$

$$F_I = m \frac{\partial^2 x}{\partial t^2} \text{ or } F_I = m\ddot{x} \tag{2.1}$$

2.2 Stiffness Component

Stiffness is the body's resistance to deformation. A deformed stiff body can restore its original shape when external force stops exerting on it. Stiffness is the most concerned component in structural design because it ensures structural stability. There are four factors affecting the stiffness of a structural member: its modulus of elasticity, shape, length and support condition.

Modulus of elasticity or Young's modulus describes the response of solid materials in the form of deformation when subjected to forces, through stress–strain relationship. Solid is a material with the highest rigidity. Strong intermolecular attraction, or internal force between solid molecules and low molecular energy that restrains the molecular movement, resulting in the ability of solid to hold its own shape. Materials react to the same applied force with different deformations because of the variation in molecular configuration among them, and thus the strength of internal force.

$$E = \frac{\sigma}{\varepsilon}$$

Stress, σ is defined as force per area, while strain, ε is the ratio of deformation to original length. Therefore, modulus of elasticity, E can be written as:

$$E = \frac{F}{A} \cdot \frac{L}{\delta}$$

where F is force, A is cross-sectional area, L is original length and δ is deformation (Fig. 2.3).

Member's shape is the key property for its resistance against bending. This property is known as the second moment of area, describes the distribution of member molecules about a specific bending axis. Bending results in rotation of cross-sections throughout the member, and resistance to bending is developed by the molecules. Molecules occupy space and they are being held together by intermolecular force. Therefore, before a cross-section can rotate, it needs enough force to overcome the attractive force and displace the molecules in contact. Like a lever, if a massive number of molecules distribute over the furthest point from the member's bending axis,

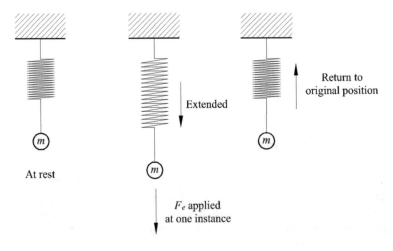

Fig. 2.3 Spring under tension

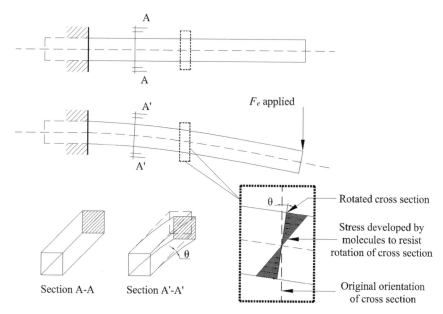

Fig. 2.4 Beam under bending

then greater force is required to rotate the cross-sections. The member is then said to have high bending resistance (Fig. 2.4).

The third factor is the length. This factor is closely related to modulus of elasticity. When a member is subject to external force, internal stress developed, and the deformation took place. Since deformation is expressed in strain, therefore it can be concluded that the magnitude of deformation is directly proportional to the original length of member. When subject to the same force, the longer member will deform more than shorter member of the same material (Fig. 2.5).

The fourth factor is the support condition of the member. In structural engineering, three most established support conditions are roller, pin and fixed. Both roller and pin do not restraint the member joint from bending, therefore the member can deform significantly on bending and is said to have low stiffness. Fixed support on the other hand, restraint the member joint from bending, therefore, the member will experience minimum deformation and is said to have high stiffness (Fig. 2.6).

Stiffness component is also known as the spring component, for its resemblance to spring's behaviour is subjected to force. When a structure is subject to force, the members are not in equilibrium as the exerted force is greater than the internal force. The work is done on the structure and it moves the members, result in deformation. By deforming, kinetic energy is converted to elastic strain energy, a form of potential energy. If there is no more external force exerted on it, the structure will stop deform after such energy conversion is completed. The elastic strain energy is then exerting

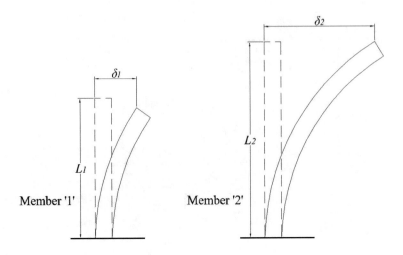

Strain, $\varepsilon = \delta/L$
Member '1' and '2' are of same material, thus $\varepsilon = \delta_1/L_1 = \delta_2/L_2$
Since $L_2 > L_1$, then $\delta_2 > \delta_1$

Fig. 2.5 Deformation of two member in different length

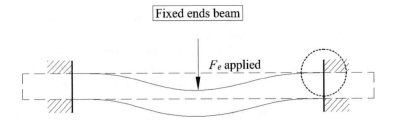

Fixed ends provide stiff restraint against rotation at support, thus reduces
the deformation (rotation and translation)

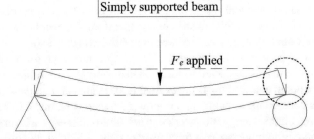

Pin and roller do not provide restraint against rotation at support, thus the
deformation is greater than fixed ends (rotation and translation)

Fig. 2.6 Beams deflection

elastic restoring force, which pull/push the members back to their original position since the original position is the most stable arrangement for the structure. Work is done again and kinetic energy increases while elastic strain energy decreases.

Stiffness component responds to the force with displacement. The force is known as a force of stiffness, F_S and it is expressed as a function of displacement, x with spring constant, k as per Hooke's law:

$$F_S = kx \tag{2.2}$$

2.3 Damping Component

When a body is subject to force, it moves because the movement is the way mass responds to force according to Newton's second law of motion. Consider a spring–mass system. It oscillates after a force is exerted on it. However, even though there is no other apparent force acting on the system, the system still oscillates with decreasing velocity and displacement. This is due to friction resulting from contact with the surrounding fluid. Consider a spring–mass system as shown in Fig. 2.7. Air resistance develops along the pendulum oscillation path and causes the amplitude of oscillation decreases with time. This phenomenon is called damping.

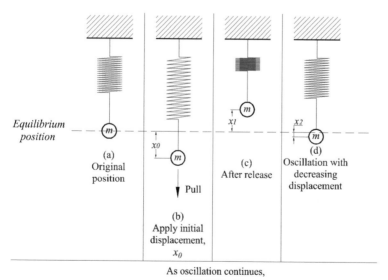

Fig. 2.7 Damping of spring-mass system

A moving structure develops damping in several ways. The first way is through friction between members. This kind of friction is developed from the joints between members. When subjected to force, structural members react differently because they have different stiffness in every case. For example, when lateral load is applied to the structure, the tendency for column to deform in the form of translation and rotation is higher than that for slab, as slab has very high rigidity in this case. Differential deformation, especially in the form of rotation occurs at the joint (the contact surface) when the members are moving in either different velocity or different direction. Dynamic friction developed as a result and converts the work done on structure to heat energy. Figure 2.8 shows an example of differential rotation at beam–column joint.

The second way of damping development is through internal friction of members. This kind of friction is significant in heterogeneous material, e.g. concrete, as its constituent materials are not evenly mixed and result in a lot of contacting surface between them. For homogeneous material, e.g. steel, all constituent materials are evenly mixed in molecular level, and the internal friction developed is negligible. When subjected to force, concrete structural member, deforms and internally, all constituent materials react and move differently. The constituent materials with differential movement will develop friction at their contact surface. Dynamic friction developed in this case as well (Fig. 2.9).

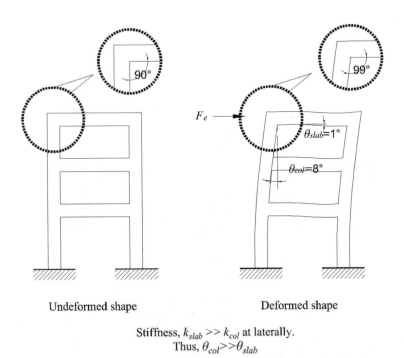

Stiffness, $k_{slab} \gg k_{col}$ at laterally.
Thus, $\theta_{col} \gg \theta_{slab}$

Fig. 2.8 Rotation at joint in frame

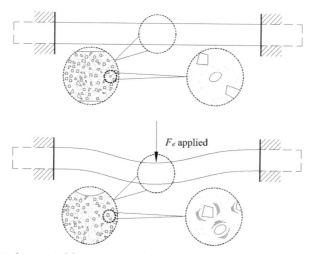

Constituent materials: aggregate, sand and cement moves upon exertion of force.
Movement like this in a closely packed arrangement develops friction between the contact
surfaces of all constituent materials.

Fig. 2.9 Single concrete member response to movement

The third way of damping development is through the damage in structural members. This is a combined effect of both stiffness and damping components. When subjected to force, structural member deforms and converts the work done to elastic strain energy. The structural member is still below its elastic limit and the resistance is mainly developed by stiffness component (elastic restoring force). If the external force persists, the member will enter a plastic state and develop more and more irreversible deformation. At a microscopic level, damage of member is caused by break of bond between particles. To overcome the attraction force and eventually break the bond, a significant amount of energy is required. Thus, by having damage in structure, energy transferred to the structure can be dissipated.

Damping can also be developed by installing a supplementary damping device on the structure. For example, a viscous damper dissipates movement (kinetic energy) by pressing the fluid inside the system. Fluid acts as resistance and develops higher damping component (Fig. 2.10).

Structure damping responds to the force with velocity. The force is known as damping force, F_D and it is expressed as a function of velocity, v with damping coefficient, C:

$$F_D = Cv$$

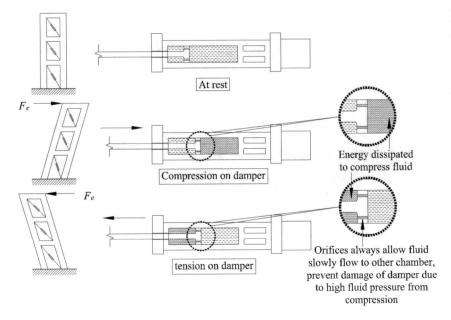

F_e

Energy dissipated
to compress fluid

Orifices always allow fluid
slowly flow to other chamber,
prevent damage of damper due
to high fluid pressure from
compression

Fig. 2.10 A functioning viscous damper

Simplify velocity, v as a differential of displacement with respect to time,

$$F_D = C \frac{\partial x}{\partial t} \text{ or } F_D = C\dot{x} \tag{2.3}$$

2.4 Equilibrium

All the above-stated components act as a reaction when dynamic force, F_e is acting
on the structure. This force is time-dependent and often acts for small amount of
time. The change in magnitude and direction of force is also rapid. As a result,
dynamic response for these applied forces on the structure is also time-dependent
and hence varies with time, due to the nature of dynamic force. The dynamic response
of structure being applied by such a dynamic force will be in terms of deformation
(displacements or rotations), velocity and acceleration.

Let's take an at rest structure as an example, as shown in Fig. 2.11. When a lateral
dynamic force is first applied to the structure, the structure generates inertial force
in response. Governing by Newton's first law of motion, work done induces kinetic
energy and moves the structure right after point O.

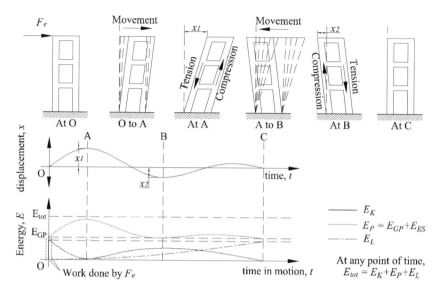

*E_{tot} = Total energy in the system; E_K = Kinetic energy; E_P = Potential energy; E_L = Energy loss in form of heat, sound etc. ; E_{GP} = Gravitational potential energy; E_{ES} = Elastic strain energy

Fig. 2.11 Structural motion as response to external force

After the force successfully pushes the structure, the columns are in tension at the side of force exertion and compression at the other side. From point O to A, the displacement continues while the columns are being extended at one side and pressed at the other side simultaneously. More and more kinetic energy has been converted into elastic strain energy.

At the same time, friction developed throughout the structure during displacement at a joint and inside the members itself. At point A, the displacement stops when kinetic energy by the mass component is fully dissipated by friction or converted to elastic strain energy.

At point B, the columns exert elastic restoring force because they want to return to their original position. Again, this force is exerted on mass component and it induced kinetic energy. Whenever there is a movement, damping component will continuously develop friction and thus, it is observable that the structure's lateral displacement is decreasing over time until the motion is fully stopped.

Therefore, the dynamic equilibrium equation is expressed as follows:

$$F_I + F_D + F_S = F_E(t)$$

Substituting Eqs. (2.1), (2.2) and (2.3) into the above equation gives us the follows:

$$m\ddot{x}(t) + C\dot{x}(t) + kx(t) = F_E(t) \tag{2.4}$$

Alternatively, express the term in the form of partial derivatives:

$$m\frac{\partial^2 x}{\partial t^2} + C\frac{\partial x}{\partial t} + kx = F_E(t) \tag{2.5}$$

Among the components, damping is the most important one in structural dynamic system because it usually dissipates 50–70% of the structural movement. Both mass and stiffness components store induce kinetic energy and elastic strain energy, respectively, forms of energy that contribute to movement. Meanwhile, damping dissipates kinetic energy by converting it to heat energy, sound energy, etc., or rather being consumed to break the bond between particles, which do not contribute to structural movement.

2.5 Idealized Structure

The three components of a dynamic system are represented in the following symbols (Fig. 2.12).

m is the total dead load on the structure/storey in kg. Dead load is weight of all objects on rigid diaphragm—slab. This includes self-weight of slab and beam, finishes and partitions, etc. C is the damping coefficient of the structure, which is commonly expressed as percentage of critical damping, C_{cr}. There is no numerical solution for the damping coefficient so far. It can only be determined through experiments. k is the stiffness of all vertical structural members. Configuration of vertical structural members can be categorized into series and parallel systems, which influences the overall stiffness of structure.

In a series system, members are connected to each other, and mass is connected to one member only. Force is only applied on that member, but it will be transmitted throughout the system via the connections. Therefore, all members arranged in series will experience the same force. Moreover, the displacement in a series system is the sum of displacement in each individual member (Fig. 2.13).

From the figure above,

$$F_{s,1} = k_1 x_1 \text{ for member '1'}$$

Fig. 2.12 Symbol of dynamic system components

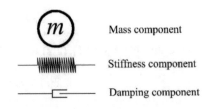

Mass component

Stiffness component

Damping component

Fig. 2.13 Series system

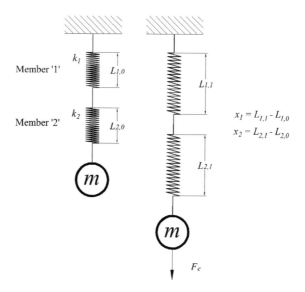

$$F_{s,2} = k_2 x_2 \text{ for member } '2'$$

Since both members experienced the same force,

$$F_s = F_{s,1} = F_{s,2}$$

Displacement in the system is the sum of displacement in each individual spring,

$$x = x_1 + x_2$$

By substituting $x_1 = \frac{F_{s,1}}{k_1}$ and $x_2 = \frac{F_{s,2}}{k_2}$ into the equation above,

$$x = \frac{F_{s,1}}{k_1} + \frac{F_{s,2}}{k_2}$$

Since $F_s = F_{s,1} = F_{s,2}$,

$$x = \frac{F_s}{k_1} + \frac{F_s}{k_2}$$

Simplify the equation above and we get,

$$x = F_s \left(\frac{1}{k_1} + \frac{1}{k_2} \right)$$

By rewriting Eq. (2.2) gives us:

$$x = F_s \times \frac{1}{k}$$

Compare the spring constant, k in both equations,

$$\frac{1}{k} = \frac{1}{k_1} + \frac{1}{k_2}$$

Thus, k for series system can be determined through:

$$\frac{1}{k} = \frac{1}{k_1} + \frac{1}{k_2} + \cdots + \frac{1}{k_n} = \sum_{i=1}^{n} \frac{1}{k_i}$$

In a parallel system, all members are connected to the same mass. Force applied is distributed among the members. Therefore, the sum of the reaction of all members is equal to the applied force. Moreover, all members work simultaneously in the system, and the displacement in the parallel system is the equals to the displacement in each individual member (Fig. 2.14).

From figure above,

$$F_{s,1} = k_1 x_1 \text{ for member '1'}$$

$$F_{s,2} = k_2 x_2 \text{ for member '2'}$$

Since force is distributed throughout the system,

$$F_s = F_{s,1} + F_{s,2}$$

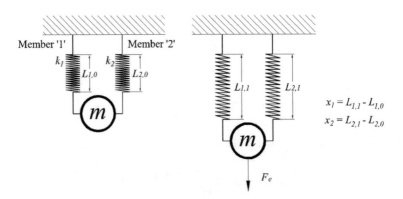

Fig. 2.14 Parallel system

Displacement in the system is equal to the displacement in each individual member,

$$x = x_1 = x_2$$

Substituting $F_{s,1} = k_1 x_1$ and $F_{s,2} = k_2 x_2$

$$F_s = k_1 x_1 + k_2 x_2$$

Since $x = x_1 = x_2$,

$$F_s = k_1 x + k_2 x$$

Simplify the equation,

$$F_s = (k_1 + k_2)x$$

Compare the spring constant, k with Eq. (2.2),

$$k = k_1 + k_2$$

Thus, k for parallel system can be determined through:

$$k = k_1 + k_2 + \cdots + k_n = \sum_{i=1}^{n} k_i$$

Example 2.1 Series and Parallel System
Identify whether the systems in Fig. 2.15 are series or parallel.

Solution

(a) A, B and C are parallel.
(b) A and B are parallel.
(c) A and B are series.
(d) A and B—AB are parallel;

   AB and C—ABC are series;
   ABC and E—$ABCE$ are parallel;
   $ABCE$ and D are series.

Example 2.2 Stiffness Component
Determine the stiffness of beam as shown in Fig. 2.16.

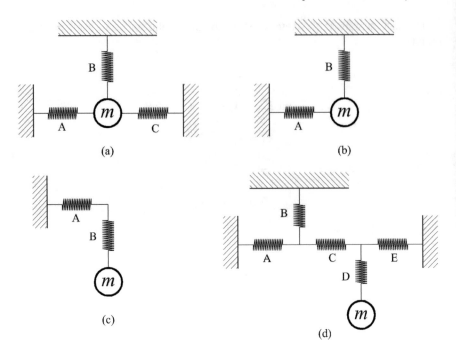

Fig. 2.15 Example 2.1

Fig. 2.16 Example 2.2

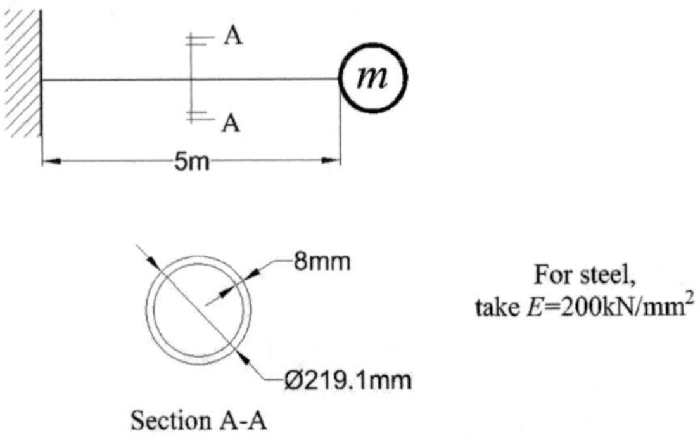

Section A-A

For steel,
take E=200kN/mm^2

Solution

By referring to Table A.1 in Appendix, second moment of area, I for circular hollow section can be determined using the following equation:

$$I = \frac{\pi d^4}{64} - \frac{\pi (d - 2t)^4}{64}$$

$$= \frac{\pi (0.2191)^4}{64} - \frac{\pi (0.2191 - 2 \times 0.008)^4}{64}$$

Fig. 2.17 Example 2.3

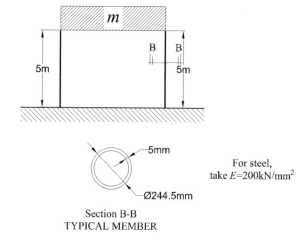

Section B-B
TYPICAL MEMBER

For steel,
take $E=200\text{kN/mm}^2$

$$= 2.96 \times 10^{-5}\,\text{m}^4$$

By referring to Table A.2 in Appendix, stiffness, k for cantilever with point load at free end can be determined using the following equation:

$$k = \frac{3EI}{L^3} = \frac{3 \times 200 \times 10^9 \times 2.96 \times 10^{-5}}{5^3} = 142,080\,\text{N/m}$$

Example 2.3 Resultant Stiffness
Determine the type of system and determine the resultant stiffness of columns for structure as shown in Fig. 2.17.

Solution
By referring to Table A.1 in Appendix, second moment of area, I for circular hollow section can be determined using the following equation:

$$
\begin{aligned}
I &= \frac{\pi d^4}{64} - \frac{\pi (d - 2t)^4}{64} \\
&= \frac{\pi (0.2445)^4}{64} - \frac{\pi (0.2445 - 2 \times 0.005)^4}{64} \\
&= 2.70 \times 10^{-5}\,\text{m}^4
\end{aligned}
$$

Both members are series. Therefore, the resultant stiffness, k of the system is as follows:

$$k = \sum_{i=1}^{n} k_i$$

By referring to Table A.2 in Appendix, stiffness for the fixed end column can be determined using the following equation:

$$k = \frac{12EI}{L^3}$$

Since both columns are typical, the following stiffness calculation applies to all of them:

$$k_1 = \frac{12 \times 200 \times 10^9 \times 2.70 \times 10^{-5}}{5^3} = 518{,}400 \, \text{N/m}$$

Resultant stiffness of the system is:

$$k = k_1 + k_2$$

Since both columns are the same,

$$k = 2k_1 = 2 \times 518{,}400 \, \text{N/m} = 1{,}036{,}800 \, \text{N/m}$$

Example 2.4 Idealized Structure
Express the structure in Fig. 2.18 in idealized form.

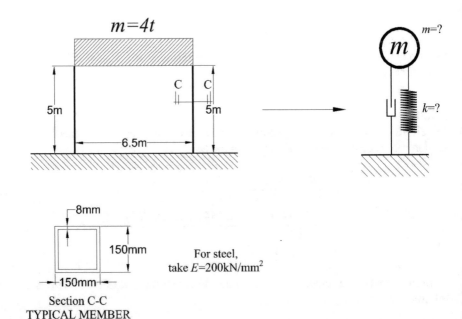

Section C-C
TYPICAL MEMBER

Fig. 2.18 Example 2.4

Solution

Mass, m is 4000 kg.

By referring to Table A.1 in Appendix, second moment of area, I for rectangular hollow section can be determined using the following equation:

$$I = \frac{bh^3}{12} - \frac{(b - 2t_w)(h - 2t_f)^3}{12}$$

$$= \frac{015 \times 0.15^3}{12} - \frac{(0.15 - 2 \times 0.008)(0.15 - 2 \times 0.008)^3}{12}$$

$$= 1.53 \times 10^{-5} \, \text{m}^4$$

Both members are parallel. Therefore, the resultant stiffness, k of the system is as follows:

$$k = \sum_{i=1}^{n} k_i$$

By referring to Table A.2 in Appendix, stiffness for the fixed end column can be determined using the following equation:

$$k = \frac{12EI}{L^3}$$

Since both columns are typical, the following stiffness calculation applies to all of them:

$$k_1 = \frac{12 \times 200 \times 10^9 \times 1.53 \times 10^{-5}}{5^3} = 293,760 \, \text{N/m}$$

Resultant stiffness of the system is:

$$k = k_1 + k_2$$

Since both columns are the same,

$$k = 2k_1 = 2 \times 293,760 \, \text{N/m} = 587,520 \, \text{N/m}$$

The idealized form of structure is shown in Fig. 2.19.

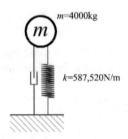

Fig. 2.19 Solution for Example 2.4

2.6   Exercises

Exercise 2.1

Identify whether the systems in Fig. 2.20 are series or parallel.

Exercise 2.2

Determine the stiffness of beam as shown in Fig. 2.21.

Exercise 2.3

Determine the type of system and determine the resultant stiffness of columns for structure as shown in Fig. 2.22.

Exercise 2.4

Express the structure in Fig. 2.23 in an idealized form.

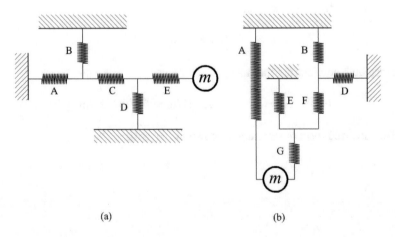

Fig. 2.20 Exercise 2.1

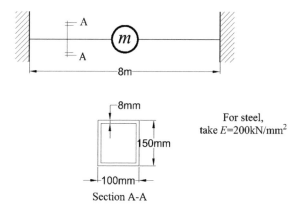

For steel,
take $E=200\text{kN/mm}^2$

Section A-A

Fig. 2.21 Exercise 2.2

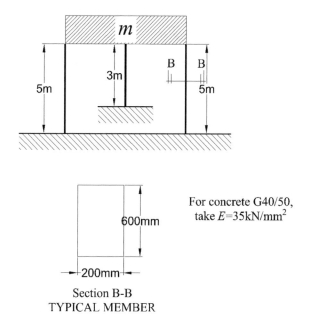

For concrete G40/50,
take $E=35\text{kN/mm}^2$

Section B-B
TYPICAL MEMBER

Fig. 2.22 Exercise 2.3

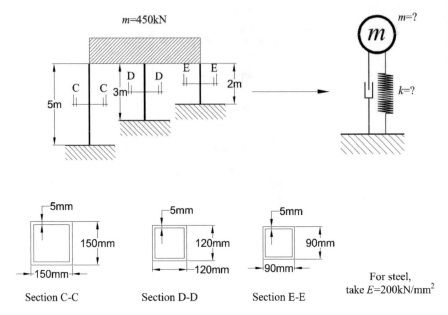

Section C-C   Section D-D   Section E-E

For steel,
take E=200kN/mm²

Fig. 2.23 Exercise 2.4

Chapter 3
Free Vibration of Single Degree of Freedom System

A single degree of freedom (SDOF) system describes structural motion with only one degree of freedom. This simplifies the analysis procedure because it is a system with only one displacement coordinate. SDOF system is often used for approximation during the early engineering stage, rather than during the detailed engineering stage (Fig. 3.1).

One of the real-life examples of SDOF is a building equipped with a base isolation system. This kind of building can be expressed in an idealized form where the entire building is the mass component, while the lateral stiffness and damping capacity of the base isolation system are stiffness and damping components respectively (Fig. 3.2).

3.1 Free Vibration of Undamped System

Free vibration is the form of vibration induced without external force, $F_E(t)$ applied on the structure. Also, the frictional forces or damping, C is neglected since the system is undamped (Fig. 3.3).

Thus, the equation of motion (Eq. 2.4) will be reduced to:

$$m\ddot{x}(t) + kx(t) = 0 \tag{3.1}$$

Equation (3.1) is homogenous equation and hence the answer of variable x complementary function. The trial solution of this second-order differential equation is trial solution which is equal to $x = Ae^{\lambda t}$. The motion of this kind of system is governed by the effect of the initial conditions as illustrated below:

© The Editor(s) (if applicable) and The Author(s), under exclusive license to Springer Nature Singapore Pte Ltd. 2020
F. Hejazi and T. K. Chun, *Conceptual Theories in Structural Dynamics*, Advanced Structured Materials 135,
https://doi.org/10.1007/978-981-15-5440-7_3

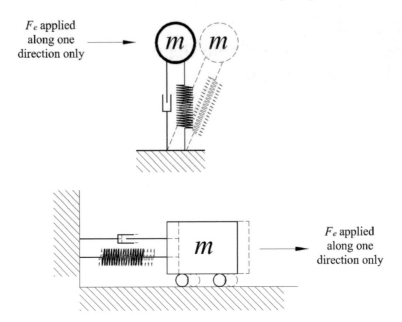

F_e applied
along one
direction only

F_e applied
along one
direction only

Fig. 3.1 Analytical models for SDOF system

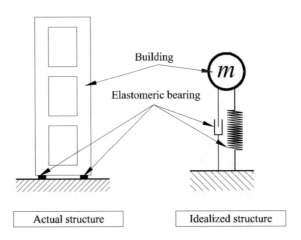

Building

Elastomeric bearing

Actual structure

Idealized structure

Fig. 3.2 Idealization of structure with a base isolation system

$$x_0 = x(0)$$
$$\dot{x}_0 = \dot{x}(0)$$
$$\ddot{x}_0 = \ddot{x}(0)$$

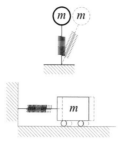

Fig. 3.3 Analytical models for undamped SDOF system under free vibration

Express Eq. (3.1) in the form of partial derivatives:

$$m\frac{\partial^2 x}{\partial t^2} + kx = 0 \tag{3.2}$$

Trial solution and its partial derivative with respect to time, t is as shown:

$$x = Ae^{\lambda t} \tag{3.3}$$

$$\dot{x} = \frac{\partial x}{\partial t} = A\lambda e^{\lambda t} \tag{3.4}$$

$$\ddot{x} = \frac{\partial^2 x}{\partial t^2} = A\lambda^2 e^{\lambda t} \tag{3.5}$$

By substituting the terms in Eqs. (3.3) and (3.5) to Eq. (3.2):

$$mA\lambda^2 e^{\lambda t} + kAe^{\lambda t} = 0$$

Simplify the equation above yields:

$$Ae^{\lambda t}\left(m\lambda^2 + k\right) = 0$$

In the form of $ab = 0$, it is either $a = 0$ or $b = 0$ to make the equation valid. In the equation above, however, $Ae^{\lambda t} \neq 0$. Therefore:

$$m\lambda^2 + k = 0$$

Express the term λ as a function of m and k:

$$\lambda^2 = -\frac{k}{m}$$

To solve λ, $i = \sqrt{-1}$ is introduced:

$$\lambda = \pm\sqrt{\frac{k}{m}}i$$

By introducing $\omega_\eta = \sqrt{\frac{k}{m}}$ to equation above:

$$\lambda_{1,2} = \pm i\omega_\eta$$

There are two possible answers for λ:

$$\lambda_1 = i\omega_\eta$$
$$\lambda_2 = -i\omega_\eta$$

By substituting the possible answers of λ into Eq. (3.3):

$$x = A_1 e^{i\omega_\eta t} + A_2 e^{-i\omega_\eta t} \tag{3.6}$$

Based on Euler's formula,

$$e^{iax} = \cos ax + i \sin ax$$
$$e^{-iax} = \cos ax - i \sin ax$$

By applying Euler's formula to Eq. (3.6) yields:

$$x = A_1(\cos \omega_\eta t + i \sin \omega_\eta t) + A_2(\cos \omega_\eta t - i \sin \omega_\eta t)$$

By rearranging the terms yields:

$$x = (A_1 + A_2) \cos \omega_\eta t + (A_1 - A_2)i \sin \omega_\eta t$$

Let $C_1 = A_1 + A_2$ and $C_2 = (A_1 - A_2)i$:

$$x = C_1 \cos \omega_\eta t + C_2 \sin \omega_\eta t \tag{3.7}$$

C_1 and C_2 are constants which can be determined through the initial conditions.

At $t = 0$, $x = x_0$. By substituting this into equation above yields:

$$x_0 = C_1 \cos(\omega_\eta \times 0) + C_2 i \sin(\omega_\eta \times 0)$$

$$C_1 = x_0 \tag{3.8}$$

Determine the partial derivatives of Eq. (3.7) with respect to time yields:

$$\dot{x} = \frac{\partial}{\partial t}(C_1 \cos \omega_\eta t + C_2 \sin \omega_\eta t) = -C_1 \omega_\eta \sin \omega_\eta t + C_2 \omega_\eta \cos \omega_\eta t$$

At $t = 0$, $\dot{x} = \dot{x}_0$. By substituting this into equation above yields:

$$\dot{x}_0 = -C_1 \omega_\eta \sin(\omega_\eta \times 0) + C_2 \omega_\eta \cos(\omega_\eta \times 0)$$

$$C_2 = \frac{\dot{x}_0}{\omega_\eta} \tag{3.9}$$

By substituting answer in Eqs. (3.8) and (3.9) into Eq. (3.7) yields:

$$x = x_0 \cos \omega_\eta t + \frac{\dot{x}_0}{\omega_\eta} \sin \omega_\eta t \tag{3.10}$$

By differentiating the equation above with respect to time yields expression for velocity, $\dot{x}$:

$$\dot{x} = -x_0 \omega_\eta \sin \omega_\eta t + \dot{x}_0 \cos \omega_\eta t \tag{3.11}$$

By differentiating the equation above with respect to time again yields expression for acceleration, $\ddot{x}$:

$$\ddot{x} = -x_0 \omega_\eta^2 \cos \omega_\eta t - \dot{x}_0 \omega_\eta \sin \omega_\eta t$$

$$\ddot{x} = -\omega_\eta^2 \left(x_0 \cos \omega_\eta t + \frac{\dot{x}_0}{\omega_\eta} \sin \omega_\eta t \right)$$

$$\ddot{x} = -\omega_\eta^2 x \tag{3.12}$$

By substituting Eqs. (3.10) and (3.12) into Eq. (3.1) yields:

$$m\left(-x_0 \omega_\eta^2 \cos \omega_\eta t - \dot{x}_0 \omega_\eta \sin \omega_\eta t\right) + k\left(x_0 \cos \omega_\eta t + \frac{\dot{x}_0}{\omega_\eta} \sin \omega_\eta t \right) = 0$$

$$-m\omega_\eta^2 \left(x_0 \cos \omega_\eta t + \frac{\dot{x}_0}{\omega_\eta} \sin \omega_\eta t \right) + k\left(x_0 \cos \omega_\eta t + \frac{\dot{x}_0}{\omega_\eta} \sin \omega_\eta t \right) = 0$$

Factorize equation above with the common term $x_0 \cos \omega_\eta t + \frac{\dot{x}_0}{\omega_\eta} \sin \omega_\eta t$ and eliminate it subsequently:

$$-m\omega_\eta^2 + k = 0 \tag{3.13}$$

$$\omega_\eta^2 = \frac{k}{m}$$

$$\omega_\eta = \sqrt{\frac{k}{m}} \tag{3.14}$$

ω_η is angular frequency of undamped system, also known as the natural angular frequency with a unit of rad/s. Period, T, which is defined as time taken for one complete oscillation in units can be expressed in terms of ω_η as follows:

$$T = \frac{2\pi}{\omega_\eta} \tag{3.15}$$

Frequency, f, which is defined as number of oscillations completed per second in unit Hz can be expressed as follows:

$$f = \frac{1}{T} = \frac{\omega_\eta}{2\pi}$$

Figure 3.4 shows typical pattern of displacement response for undamped system under free vibration. Without damping, the amplitude of vibration will not be decreased. The energy is conserved within the system, only being converted from potential energy to kinetic energy and vice versa.

Example 3.1 Free Vibration of Undamped System

A steel frame shown in Fig. 3.5 is fixed at the base has horizontal rigid member with mass of 2 ton/m. The frame columns are made of steel hollow section with

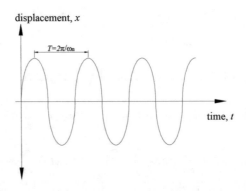

Fig. 3.4 Displacement response of undamped system under free vibration

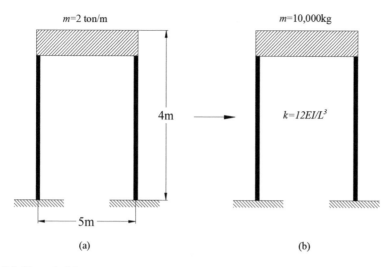

Fig. 3.5 Example 3.1

dimensions of 100×100 mm and thickness of 10 mm. If the frame was subjected to free vibration with initial displacement of 60 mm and initial velocity of 0.02 m/s.

(a) Determine the dynamic response of this frame.
(b) Determine the maximum displacement, velocity and acceleration.

Solution
By referring to Table A.1 in Appendix, second moment of area, I for rectangular hollow section can be determined using the following equation:

$$I = \frac{bh^3}{12} - \frac{(b - 2t_w)(h - 2t_f)^3}{12} = \frac{0.1^4 - 0.08^4}{12} = 4.92 \times 10^{-6}\,\mathrm{m}^4$$

By referring to Table A.2 in Appendix, stiffness, k for two fixed ends columns in parallel can be determined using the following equation:

$$k = 2 \times \frac{12EI}{L^3} = \frac{2 \times 12 \times 200 \times 10^9 \times 4.92 \times 10^{-6}}{4^3} = 369,000\,\mathrm{N/m}$$

The natural frequency of the structure in this case is:

$$\omega_n = \sqrt{\frac{k}{m}} = \sqrt{\frac{369,000}{10,000}} = 6.075 \, \text{rad/s}$$

The natural period and frequency (in Hz) of the structure in this case is:

$$T = \frac{2\pi}{\omega_n} = \frac{2\pi}{6.075} = 1.034 \, \text{s}$$

$$f = \frac{1}{T} = \frac{1}{1.034} = 0.97 \, \text{Hz}$$

(a) Dynamic Response

The dynamic response of undamped frame system under free vibration can be evaluated through the displacement, velocity and acceleration as follow:
Displacement Response:

$$x = x_0 \cos \omega_n t + \frac{\dot{x}_0}{\omega_n} \sin \omega_n t = 0.06 \cos(6.075t) + \frac{0.02}{6.075} \sin(6.075t)$$

Velocity Response:

$$\dot{x} = -x_0 \omega_n \sin \omega_n t + \dot{x}_0 \cos \omega_n t = -0.365 \sin(6.075t) + 0.02 \cos(6.075t)$$

Acceleration Response:

$$\ddot{x} = -x_0 \omega_n^2 \cos \omega_n t - \dot{x}_0 \omega_n \sin \omega_n t$$

$$= -\omega_n^2 \left(x_0 \cos \omega_n t + \frac{\dot{x}_0}{\omega_n} \sin \omega_n t \right) = -\omega_n^2 x$$

$$= -6.075^2 \left(0.06 \cos(6.075t) + \frac{0.02}{6.075} \sin(6.075t) \right)$$

$$= -2.214 \cos(6.075t) - 0.122 \sin(6.075t)$$

Figure 3.6 shows the response plot for the structure in this case:

(b) Maximum dynamic response

$$x_m = \sqrt{x_o^2 + \left(\frac{\dot{x}_o}{\omega_n} \right)^2} = \sqrt{0.06^2 + \left(\frac{0.02}{6.075} \right)^2} = 0.06$$

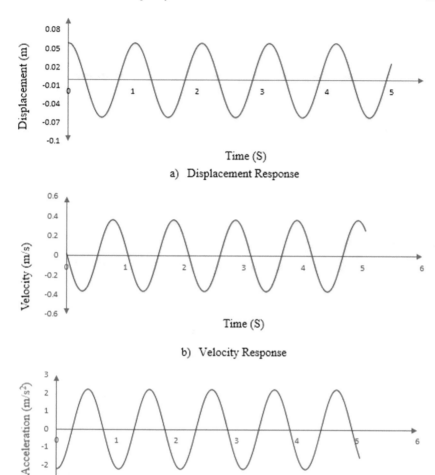

Fig. 3.6 Displacement, velocity and acceleration response for the frame in Example 3.1

Maximum Displacement

$$x_{max} = x_m = 0.06 \, \text{m}$$

Maximum Velocity

$$\dot{x}_{max} = x_m \omega_n = 0.06 \times 6.075 = 0.365 \, \text{m/s}$$

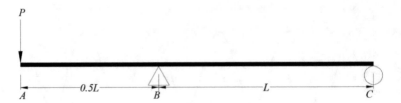

Fig. 3.7 Example 3.2

Maximum Acceleration

$$\ddot{x}_{max} = x_m \omega_n^2 = 0.06 \times 6.075^2 = 2.214 \, \text{m/s}$$

Example 3.2 Free Vibration of Undamped System

Determine the natural period and frequency of beam shown in Fig. 3.7.

Solution

$$\sum M_C = 0$$
$$P(1.5L) - R_B(L) = 0$$
$$R_B = 1.5P$$
$$\sum F_y = 0$$
$$-P + 1.5P + R_C = 0$$
$$R_C = -0.5P$$

By using strain energy method to determine the deflection of beam at point A

$$U = U_{AB} + U_{CB} = \int_0^{0.5L} \frac{M_{AB}^2}{2EI} + \int_0^{L} \frac{M_{CA}^2}{2EI}$$

Both integrals can be transformed to:

$$\int_0^{0.5L} \frac{M_{AB}^2}{2EI} = \int_0^{0.5L} \frac{(-PX)^2}{2EI} = \frac{P^2}{2EI} \int_0^{0.5L} x^2 dx = \frac{P^2}{2EI} \left[\frac{x^3}{3} \right]_0^{0.5L}$$

$$= \frac{P^2}{6EI} \left[\frac{L^3}{8} \right] = \frac{P^2 L^3}{48EI}$$

$$\int_0^{L} \frac{M_{CA}^2}{2EI} = \int_0^{l} \frac{(-0.5PX)^2}{2EI} = \frac{P^2}{8EI} \int_0^{L} x^2 dx = \frac{P^2}{8EI} \left[\frac{x^3}{3} \right]_0^{L}$$

$$= \frac{P^2}{8EI}\left[\frac{L^3}{3}\right] = \frac{P^2 L^3}{24EI}$$

Therefore,

$$U = U_{AB} + U_{BC} = \frac{P^2 L^3}{48EI} + \frac{P^2 L^3}{24EI} = \frac{P^2 L^3}{16EI}$$

The displacement of structure can be expressed in the following form:

$$\delta_c = \frac{\partial U}{\partial P} = \frac{\partial}{\partial P}\left(\frac{P^2 L^3}{16EI}\right) = \frac{PL^3}{8EI}$$

The stiffness can be determined using equation below:

$$k = \frac{P}{\delta_c} = \frac{8EI}{L^3}$$

The natural frequency of structure in this case is:

$$\omega_n = \sqrt{\frac{K}{m}} = \sqrt{\frac{8EI}{mL^3}}$$

The natural period of the structure in this case is:

$$T = 2\pi \sqrt{\frac{mL^3}{8EI}}$$

Example 3.3 Free Vibration of Undamped System

Determine the natural period and frequency of steel beam shown in Fig. 3.8.

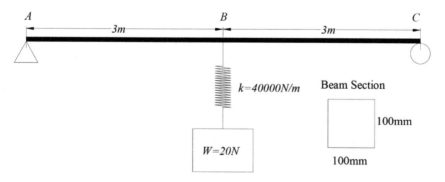

Fig. 3.8 Example 3.3

Solution

By referring to Table A.1 in Appendix, second moment of area, I for rectangular section can be determined using the following equation:

$$I = \frac{bh^3}{12} = \frac{0.1^4}{12} = 8.33 \times 10^{-6} \, m^4$$

By referring to Table A.2 in Appendix, stiffness, k for simply supported beam subjected to point load at midspan can be determined using the following equation:

$$k_{eqv} = \frac{48EI}{L^3} = \frac{48 \times 200 \times 10^9 \times 8.33 \times 10^{-6}}{6^3}$$
$$= 370,222.222 \, N/m$$

The structure is a series system, therefore:

$$\frac{1}{k} = \frac{1}{k_1} + \frac{1}{k_2} = \frac{1}{370,222.222} + \frac{1}{40,000} = 2.77 \times 10^{-5}$$

$$k = 36,099.675 \, N/m$$

The weight of load is 20 N, which can be converted to kg using the following equation:

$$W = \frac{20}{9.81} \, kg = 2.039 \, kg$$

The natural frequency of the structure in this case is:

$$\omega_n = \sqrt{\frac{k}{m}} = \sqrt{\frac{36,099.675}{2.039}} = 133.04 \, rad/s$$

The natural period and frequency (in Hz) of the structure in this case is:

$$T = \frac{2\pi}{\omega_n} = \frac{2\pi}{133.04} = 0.047 \, s$$
$$f = \frac{1}{T} = \frac{1}{0.047} = 21.2 \, Hz$$

3.2 Free Vibration of Damped System

Free vibration is the form of vibration induced without external force, $F_E(t)$ applied on the structure. Since this is a damped system, frictional forces or damping, C is not neglected (Fig. 3.9).

Therefore, the equilibrium equation (Eq. 2.4) will be reduced to:

$$m\ddot{x}(t) + C\dot{x}(t) + kx(t) = 0 \tag{3.16}$$

Equation (3.16) is homogenous equation and hence the answer of variable x complementary function. The trial solution of this second-order differential equation is trial solution which is equal to $x = Ae^{\lambda t}$ (Eq. 3.3). The motion of this kind of system is governed by the effect of the initial conditions as illustrated below:

$$x_0 = x(0)$$
$$\dot{x}_0 = \dot{x}(0)$$
$$\ddot{x}_0 = \ddot{x}(0)$$

Express Eq. (3.16) in the form of partial derivatives:

$$m\frac{\partial^2 x}{\partial t^2} + c\frac{\partial x}{\partial t} + kx = 0 \tag{3.17}$$

Substituting terms in Eqs. (3.3), (3.4) and (3.5) into Eq. (3.17) yields:

$$mA\lambda^2 e^{\lambda t} + cA\lambda e^{\lambda t} + kAe^{\lambda t} = 0$$

Simplify the equation above:

$$Ae^{\lambda t}\left(m\lambda^2 + c\lambda + k\right) = 0$$

Fig. 3.9 Analytical models for damped SDOF system under free vibration

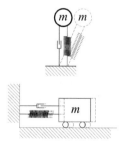

In the form of $ab = 0$, it is either $a = 0$ or $b = 0$ to make the equation valid. In the equation above, however, $Ae^{\lambda t} \neq 0$. Therefore:

$$m\lambda^2 + c\lambda + k = 0$$

Divide the terms in the equation above by m:

$$\lambda^2 + \frac{c}{m}\lambda + \frac{k}{m} = 0$$

Comparing the quadratic equation above with general quadratic equation $ax^2 + bx + c = 0$. The coefficients of the quadratic equations are identified as follows:

$$a = 1$$
$$b = \frac{c}{m}$$
$$c = \frac{k}{m}$$

The quadratic roots, λ_1 and λ_2 can be solved by using general solution $x = \frac{-b \pm \sqrt{b^2 - 4ac}}{2a}$ corresponding to the general form of quadratic equation:

$$\lambda_{1,2} = \frac{-\frac{c}{m} \pm \sqrt{\left(\frac{c}{m}\right)^2 - \frac{4k}{m}}}{2}$$

Simplify the equation above yields:

$$\lambda_{1,2} = -\frac{c}{2m} \pm \sqrt{\left(\frac{c}{2m}\right)^2 - \frac{k}{m}} \qquad (3.18)$$

There are three cases of damping behaviour of structure based on the value of discriminant, $\left(\frac{c}{2m}\right)^2 - \frac{k}{m}$.

3.2.1   Critical Damping

Critical damping occurs when $\left(\frac{c}{2m}\right)^2 - \frac{k}{m} = 0$
Therefore, from this condition we can say that:

$$\left(\frac{c}{2m}\right)^2 = \frac{k}{m}$$

Square root the terms on both sides yields:

$$\frac{c}{2m} = \sqrt{\frac{k}{m}}$$

Express c in terms of m and k:

$$c = 2m\sqrt{\frac{k}{m}}$$

Simplify the equation above yields:

$$c = 2\sqrt{km}$$

By introducing critical damping, $c_{cr} = 2\sqrt{km}$ to the equation above:

$$c = c_{cr}$$

Damping ratio, $\xi = \frac{c}{c_{cr}}$, is the ratio of structural damping to its critical damping. Under this condition, $\xi = \frac{c_{cr}}{c_{cr}} = 1$. Under this scenario, $\lambda_{1,2} = -\frac{c}{2m}$, which are two real and equal roots obtained by substituting $\left(\frac{c}{2m}\right)^2 - \frac{k}{m} = 0$ to Eq. (3.18). By substituting the roots to a trial solution, Eq. (3.3) yields:

$$x = A_1 e^{-\frac{c}{2m}} + A_2 e^{-\frac{c}{2m}}$$
$$x = (A_1 + A_2)e^{-\frac{c}{2m}}$$

Under critical damping condition, the structure will not be able to vibrate after being displaced. The structure moves toward its original position upon release, but the damping component is too strong, and it dissipates the energy quickly. Figure 3.10 shows the typical pattern of displacement response for a freely vibrating system subjected to critical damping ($\xi = 1$) condition.

3.2.2 Overdamping

Overdamping occurs when $\left(\frac{c}{2m}\right)^2 - \frac{k}{m} > 0$
Therefore, from the expression above we know that:

$$\left(\frac{c}{2m}\right)^2 > \frac{k}{m}$$

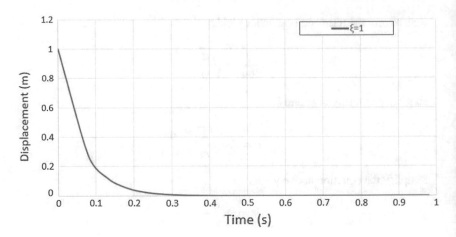

Fig. 3.10 Displacement response of damped system under free vibration (critical damping condition)

Square root the terms on both sides yields:

$$\frac{c}{2m} > \sqrt{\frac{k}{m}}$$

Express c in terms of m and k:

$$c > 2m\sqrt{\frac{k}{m}}$$

Simplify the equation above yields:

$$c > 2\sqrt{km}$$

Since critical damping, $c_{cr} = 2\sqrt{km}$:

$$c > c_{cr}$$

Therefore, damping ratio, $\xi = \frac{c}{c_{cr}} > 1$.

Under this scenario, $\lambda_{1,2} = -\frac{c}{2m} \pm \sqrt{\left(\frac{c}{2m}\right)^2 - \frac{k}{m}}$, which are two real and distinct roots. By substituting the roots to trial solution, Eq. (3.3) yields:

$$x = A_1 e^{-\frac{c}{2m}+\sqrt{\left(\frac{c}{2m}\right)^2 - \frac{k}{m}}} + A_2 e^{-\frac{c}{2m}-\sqrt{\left(\frac{c}{2m}\right)^2 - \frac{k}{m}}}$$

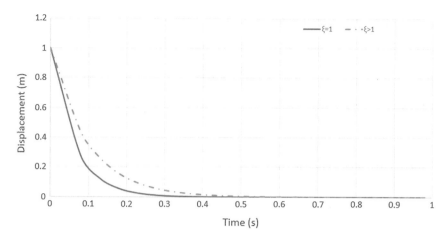

Fig. 3.11 Displacement response of damped system under free vibration (critical and overdamping condition)

Overdamping is a condition to avoid. When a structure is under such condition, the stiffness is high and when it is subjected to loading, the structure might fail abruptly without precaution. Figure 3.11 compares the typical patterns of displacement responses for a freely vibrating system subjected to critical ($\xi = 1$) and overdamping ($\xi > 1$) conditions.

3.2.3 Underdamping

Underdamping occurs when $\left(\frac{c}{2m}\right)^2 - \frac{k}{m} < 0$

Therefore, the following correlation can be obtained:

$$\left(\frac{c}{2m}\right)^2 < \frac{k}{m}$$

Square root the terms on both sides yields:

$$\frac{c}{2m} < \sqrt{\frac{k}{m}}$$

Express c in terms of m and k:

$$c < 2m\sqrt{\frac{k}{m}}$$

Simplify the equation above yields:

$$c < 2\sqrt{km}$$

Since critical damping, $c_{cr} = 2\sqrt{km}$:

$$c < c_{cr}$$

Therefore, damping ratio, $\xi = \frac{c}{c_{cr}} < 1$. Under this scenario, $\lambda_{1,2} = -\frac{c}{2m} \pm \sqrt{\left(\frac{c}{2m}\right)^2 - \frac{k}{m}}$, which are two imaginary roots. The term $\frac{c}{2m}$ can be transformed by multiplying $\frac{c_{cr}}{c_{cr}}$ to simplify the answer for λ:

$$\frac{c}{2m} = \frac{c}{2m} \times \frac{c_{cr}}{c_{cr}}$$

Rearrange equation above:

$$\frac{c}{2m} = \frac{c}{c_{cr}} \times \frac{c_{cr}}{2m}$$

Since $\xi = \frac{c}{c_r}$ and $c_r = 2\sqrt{km}$, therefore:

$$\frac{c}{2m} = \xi \times \frac{2\sqrt{km}}{2m}$$

Simplify the equation above yields:

$$\frac{c}{2m} = \xi \times \sqrt{\frac{k}{m}}$$

Since $\omega_\eta = \sqrt{\frac{k}{m}}$, the equation above can be simplified:

$$\frac{c}{2m} = \xi\omega_\eta \tag{3.19}$$

By substituting $\frac{c}{2m} = \xi\omega_\eta$ and $\omega_\eta = \sqrt{\frac{k}{m}}$ into Eq. (3.18) yields:

$$\lambda_{1,2} = -\xi\omega_n \pm \sqrt{(\xi\omega_n)^2 - \omega_n^2}$$

Factorize equation above with the common term $-\omega_n^2$ (in surd) yields:

$$\lambda_{1,2} = -\xi\omega_n \pm \sqrt{-\omega_n^2(1-\xi^2)}$$

Simplify the equation above using $i = \sqrt{-1}$ yields:

$$\lambda_{1,2} = -\xi\omega_n \pm i\omega_n\sqrt{(1-\xi^2)}$$

By defining angular frequency of damped system, $\omega_d = \omega_n\sqrt{(1-\xi^2)}$, the equation above can be simplified to:

$$\lambda_{1,2} = -\xi\omega_n \pm i\omega_d$$

By substituting the equation above into trial solution (Eq. 3.3) yields:

$$x = A_1 e^{(-\xi\omega_n + i\omega_d)t} + A_2 e^{(-\xi\omega_n - i\omega_d)t}$$

Factorize equation above with the common term $e^{-\xi\omega_n t}$ yields:

$$x = e^{-\xi\omega_n t}\left(A_1 e^{i\omega_d t} + A_2 e^{-i\omega_d t}\right)$$

Applying Euler's formula to the equation above yields:

$$x = e^{-\xi\omega_n t}[(A_1\cos\omega_d t + A_1 i\sin\omega_d t) + (A_2\cos\omega_d t - A_2 i\sin\omega_d t)]$$

Rearrange the equation above:

$$x = e^{-\xi\omega_n t}[(A_1\cos\omega_d t + A_2\cos\omega_d t) + (A_1 i\sin\omega_d t - A_2 i\sin\omega_d t)]$$

Factorize equation above with the common terms $\cos\omega_d t$ and $\sin\omega_d t$:

$$x = e^{-\xi\omega_n t}[(A_1 + A_2)\cos\omega_d t + i(A_1 - A_2)\sin\omega_d t]$$

Let $C_1 = (A_1 + A_2)$ and $C_2 = (A_1 - A_2)i$:

$$x = e^{-\xi\omega_n t}[C_1\cos\omega_d t + C_2\sin\omega_d t] \qquad (3.20)$$

C_1 and C_2 are constants which can be determined through the initial conditions. At $t = 0$, $x = x_0$. By substituting this into the equation above yields:

$$x_0 = e^{-\xi\omega_n(0)}C_1\cos(\omega_d \times 0) + C_2 i\sin(\omega_d \times 0)$$

$$C_1 = x_0 \tag{3.21}$$

Differentiation using product rule can be performed as follows:

$$y = f(x) = uv$$

$$\frac{dy}{dx} = f'(x) = v\frac{du}{dx} + u\frac{dv}{dx}$$

Determine the partial derivatives of Eq. (3.20) with respect to time using product rule yields:

$$\dot{x} = -\xi\omega_n e^{-\xi\omega_n t}(C_1 \cos\omega_d t + C_2 \sin\omega_d t)$$
$$+ e^{-\xi\omega_n t}(-C_1\omega_d \sin\omega_d t + C_2\omega_d \cos\omega_d t) \tag{3.22}$$

At $t = 0$, $\dot{x} = \dot{x}_0$. By substituting this into equation above yields:

$$\dot{x} = -\xi\omega_n e^{-\xi\omega_n(0)}(C_1 \cos\omega_d \times 0 + C_2 \sin\omega_d \times 0)$$
$$+ e^{-\xi\omega_n(0)}(-C_1\omega_d \sin\omega_d \times 0 + C_2\omega_d \cos\omega_d \times 0)$$

$$\dot{x}_o = -\xi\omega_n C_1 + C_2\omega_d$$

Since $C_1 = x_0$, the equation can be transformed into:

$$\dot{x}_o = -\xi\omega_n x_o + C_2\omega_d$$

Rearrange equation above yields:

$$C_2 = \frac{\dot{x}_o + \xi\omega_n x_o}{\omega_d} \tag{3.23}$$

By substituting (3.21) and (3.23) into Eq. (3.20) yields:

$$x = e^{-\xi\omega_n t}\left(x_o \cos\omega_d t + \frac{\dot{x}_o + \xi\omega_n x_o}{\omega_d} \sin\omega_d t\right) \tag{3.24}$$

By differentiating the equation above with respect to time yields expression for velocity, $\dot{x}$:

$$\dot{x} = -\xi\omega_n e^{-\xi\omega_n t}\left[x_o \cos\omega_d t + \frac{\dot{x}_o + \xi\omega_n x_o}{\omega_d} \sin\omega_d t\right]$$
$$+ e^{-\xi\omega_n t}[-x_o\omega_d \sin\omega_d t + (\dot{x}_o + \xi\omega_n x_o) \cos\omega_d t]$$

As defined above, angular frequency of the damped system is:

$$\omega_d = \omega_n\sqrt{\left(1 - \xi^2\right)}$$

Period for the damped system, T_d, which is defined as time taken for one complete oscillation in unit s can be expressed in terms of ω_d as follows:

$$T_d = \frac{2\pi}{\omega_d} = \frac{2\pi}{\omega_n\sqrt{\left(1 - \xi^2\right)}} \qquad (3.25)$$

Frequency for the damped system, f_d, which is defined as number of oscillations completed per second in unit Hz can be expressed as follows:

$$f_d = \frac{1}{T_d} = \frac{\omega_n\sqrt{\left(1 - \xi^2\right)}}{2\pi}$$

Figure 3.12 compares the typical patterns of displacement responses for a freely vibrating system subjected to critical ($\xi = 1$), overdamping ($\xi > 1$) and underdamping ($\xi < 1$) conditions.

Example 3.4 Free Vibration of Damped System

The steel frame shown in Fig. 3.13 has a horizontal rigid member with mass of 10 ton and it was subjected to free vibration with initial displacement of 300 mm and initial velocity of 50 mm/s. The damping ratio of frame system is assumed to 2%.

(a) Find the dynamic response of frame system in term displacement and velocity.

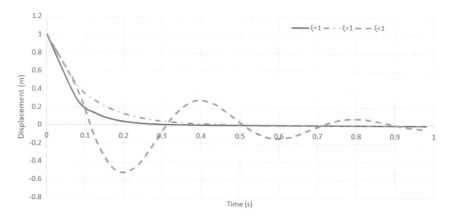

Fig. 3.12 Displacement response of damped system under free vibration

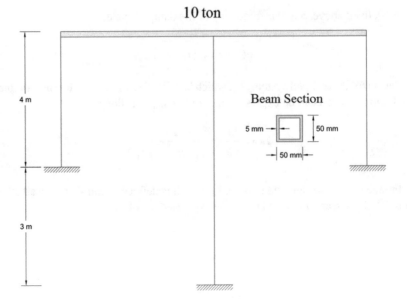

Fig. 3.13 Example 3.4

(b) Determine the displacement and velocity after 5 s.

Solution
By referring to Table A.1 in Appendix, second moment of area, I for rectangular hollow section can be determined using the following equation:

$$I = \frac{bh^3}{12} - \frac{(b - 2t_w)(h - 2t_f)^3}{12} = \frac{0.05^4 - 0.04^4}{12} = 3.075 \times 10^{-7}\,\mathrm{m^4}$$

By referring to Table A.2 in Appendix, stiffness, k for fixed end columns in parallel can be determined using the following equation:

$$k_{eqv} = 2 \times \frac{12EI}{L_1^3} + \frac{12EI}{L_2^3}$$
$$= \frac{2 \times 12 \times 200 \times 10^9 \times 3.075 \times 10^{-7}}{4^3} + \frac{12 \times 200 \times 10^9 \times 3.075 \times 10^{-7}}{7^3}$$
$$= 25,214.1\,\mathrm{N/m}$$

The natural frequency of the structure in this case is

$$\omega_n = \sqrt{\frac{k}{m}} = \sqrt{\frac{25,214.1}{10,000}} = 1.588\,\mathrm{rad/s}$$

The angular frequency of damped system is

$$\omega_d = \omega_n\sqrt{(1-\xi^2)} = 1.588 \times \sqrt{(1-0.02^2)} = 1.588\,\text{rad/s}$$

The period and frequency (in Hz) of the damped system is

$$T_d = \frac{2\pi}{\omega_d} = \frac{2\pi}{1.588} = 3.957\,\text{s}$$

$$f = \frac{1}{T} = \frac{1}{3.957} = 0.253\,\text{Hz}$$

(a) Dynamic response

The dynamic response is determined in terms of displacement, velocity and acceleration as follows:

$$\frac{\dot{x}_o + \xi\omega_n x_o}{\omega_d} = \frac{0.05 + 0.02 \times 1.59 \times 0.3}{1.59} = 0.0375$$

Displacement response

$$x = e^{-0.02\times1.588t}(0.3\cos 1.588t + 0.0375\sin 1.588t)$$
$$= e^{-0.0318t}(0.3\cos 1.588t + 0.0375\sin 1.588t)$$

Velocity response

$$\dot{x} = -0.0318 \times e^{-0.0318t}(0.3\cos 1.588t + 0.0375\sin 1.588t)$$
$$+ e^{-0.0318t}(-0.477\sin 1.588t + 0.06\cos 1.588t)$$

(b) Displacement and velocity response after 5 s

Displacement response after 5 s:

$$x = e^{-0.0318\times5}(0.3\cos 1.588 \times 5 + 0.0375\sin 1.588 \times 5)$$
$$= 0.01\,\text{m}$$

Velocity response after 5 s:

$$\dot{x} = -0.0318 \times e^{-0.0318\times5}(0.3\cos 1.588 \times 5 + 0.0375\sin 1.588 \times 5)$$
$$+ e^{-0.0318\times5}(-0.477\sin 1.588 \times 5 + 0.06\cos 1.588 \times 5)$$
$$= -0.41\,\text{m/s}$$

The dynamic response of given steel frame is illustrated in Fig. 3.14. The dynamic response is presented in terms of displacement and velocity only at damping ratio of 0.02.

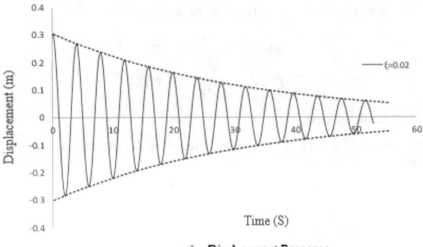

a)   Displacement Response

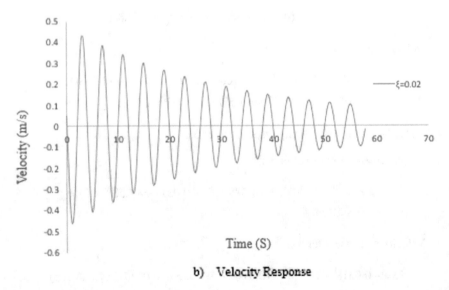

b)   Velocity Response

Fig. 3.14 Dynamic response for Example 3.4

3.3 Logarithmic Decremental

Logarithmic decrement is a method that is practically used to determine the damping ratio of the system. Figure 3.15 shows the free vibration undamped system.

At $t = t_1$ and $x = x_1$,

The function for displacement response can be written based on Eq. (3.24)

$$x_1 = e^{-\xi \omega_n t_1} [C_1 \cos \omega_d t_1 + C_2 \sin \omega_d t_1] \tag{3.26}$$

Let t_2 be the time where the subsequent cycle is completed, and thus it is a period, T_d away from t_1:

$$t_2 = T_d + t_1$$

By substituting Eq. (3.25) into equation above yields:

$$t_2 = t_1 + \frac{2\pi}{\omega_d}$$

Rewrite the equation above yields:

$$t_2 = \frac{\omega_d t_1 + 2\pi}{\omega_d} \tag{3.27}$$

The function for x_2 can be written based on Eq. (3.24)

$$x_2 = e^{-\xi \omega_n t_2} [C_1 \cos \omega_d t_2 + C_2 \sin \omega_d t_2]$$

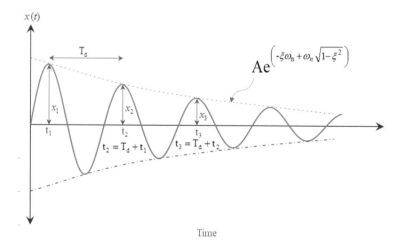

Fig. 3.15 Undamped system under free vibration

By substituting Eq. (3.27) into equation above yields:

$$x_2 = e^{-\xi\omega_n(T_d+t_1)}\left[C_1\cos\omega_d\left(\frac{\omega_d t_1 + 2\pi}{\omega_d}\right) + C_2\sin\omega_d\left(\frac{\omega_d t_1 + 2\pi}{\omega_d}\right)\right]$$

Simplify the equation above yields:

$$x_2 = e^{-\xi\omega_n(T_d+t_1)}[C_1\cos(\omega_d t_1 + 2\pi) + C_2\sin(\omega_d t_1 + 2\pi)]$$

By applying trigonometry identity $\cos(A + 2\pi) = \cos A$ and $\sin(A + 2\pi) = \sin A$ into equation above yields:

$$x_2 = e^{-\xi\omega_n(T_d+t_1)}[C_1\cos\omega_d t_1 + C_2\sin\omega_d t_1] \qquad (3.28)$$

By dividing Eq. (3.26) with Eq. (3.28) yields:

$$\frac{x_1}{x_2} = \frac{e^{-\xi\omega_n t_1}}{e^{-\xi\omega_n(T_d+t_1)}} \times \frac{[C_1\cos\omega_d t_1 + C_2\sin\omega_d t_1]}{[C_1\cos\omega_d t_1 + C_2\sin\omega_d t_1]}$$

Simplify the equation above yields:

$$\frac{x_1}{x_2} = \frac{e^{-\xi\omega_n t_1}}{e^{-\xi\omega_n(T_d+t_1)}}$$

$$\frac{x_1}{x_2} = \frac{e^{-\xi\omega_n t_1}}{e^{-\xi\omega_n T_d} \times e^{-\xi\omega_n t_1}}$$

$$\frac{x_1}{x_2} = \frac{1}{e^{-\xi\omega_n T_d}}$$

$$\frac{x_1}{x_2} = e^{\xi\omega_n T_d}$$

By substituting Eq. (3.25) into equation above yields:

$$\frac{x_1}{x_2} = e^{\frac{2\pi\xi}{\sqrt{1-\xi^2}}}$$

Let $\delta = \ln\left(\frac{x_1}{x_2}\right)$:

$$\delta = \frac{2\pi\xi}{\sqrt{1-\xi^2}}$$

If the damping ratio, ξ is small

$$\delta = 2\pi\xi$$

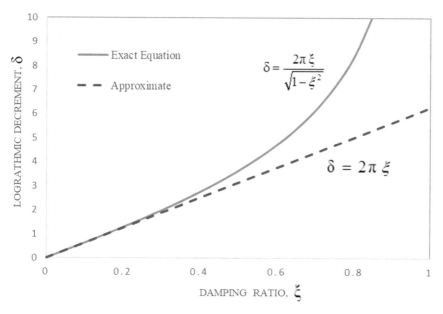

Fig. 3.16 Exact and approximated solution of logarithmic decrement

Rearrange the equation above yields:

$$\delta\sqrt{1-\xi^2} = 2\pi\xi$$

Square the terms on both sides to eliminate the surd yields:

$$\delta^2 - \delta^2\xi^2 = 4\pi^2\xi^2$$

Rearrange the equation above yields:

$$\delta^2 = 4\pi^2\xi^2 + \delta^2\xi^2$$

Simplify the equation above yields:

$$\delta^2 = \xi^2\left(4\pi^2 + \delta^2\right)$$

Rearrange the equation above yields:

$$\xi^2 = \frac{\delta^2}{4\pi^2 + \delta^2}$$

Figure 3.16 shows logarithmic decrement response due to the exact and the approximation equation. The approximated solution converges with the exact solution until $\xi \geq 30\%$.

The following relationships can be applied to evaluate the displacement response over several cycles:

$$\frac{x_0}{x_N} = \frac{x_0}{x_1} \cdot \frac{x_1}{x_2} \cdots \cdots \frac{x_{N-1}}{x_N} = \left(e^{\xi \omega_n T_d}\right)^N = e^{\xi \omega_n T_d N}$$

At the Nth cycle,

$$\delta = \frac{1}{N} \ln\left(\frac{x_0}{x_N}\right)$$

$$\delta = \xi \omega_n T_d N$$

$$\delta = \frac{2\pi N \xi}{\sqrt{1 - \xi^2}}$$

If the damping ratio, ξ is small,

$$\delta \cong 2\pi N \xi$$

Also, the damping ratio can be determined using equation below:

$$\xi = \frac{1}{2\pi N} \delta$$

Example 3.5 Logarithmic Decremental

A water tank as shown in Fig. 3.17 is subjected to lateral load of 200 kN that causes maximum lateral displacement of 10 mm. (a) Determine the damping of the structure if the damping ratio $\xi = 5\%$. (b) How long does it take to reach 2 mm displacement?

Solution

(a) Damping of structure system can be determined using $\xi = \frac{C}{C_{cr}}$ and $C_{cr} = 2\sqrt{km}$

$$k = \frac{F}{\delta} = \frac{200 \times 10^3}{0.01} = 2 \times 10^7 \, \text{N/m}$$

$$C = 2\xi \sqrt{km} = 2 \times 0.05 \times \sqrt{2 \times 10^7 \times 5 \times 10^3}$$
$$= 31622.77 \, \text{N} \cdot \text{s/m}$$

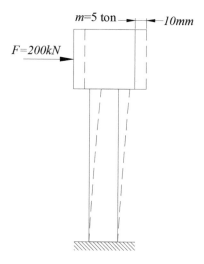

Fig. 3.17 Example 3.5

(b) Time required to reach 2 mm of lateral displacement

Number of cycles need to reach 2 mm displacement can be determined through logarithmic decrement. Since damping ratio is small, the approximate solution is implemented here.

$$\ln\left(\frac{x_1}{x_2}\right) = 2\pi N\xi \ln\left(\frac{x_1}{x_2}\right) = \ln\left(\frac{0.01}{0.002}\right) = 1.609$$

$$N = \frac{1.609}{2\pi\xi} = \frac{1.609}{2\pi \times 0.05} = 5.12\,\text{cycle}$$

The damped time period per one cycle is required to be determined:

$$\omega_n = \sqrt{\frac{K}{m}} = \sqrt{\frac{2 \times 10^7}{5000}} = 63.25\,\text{rad/s}$$

$$\omega_d = \omega_n\sqrt{(1 - \xi^2)} = 63.25\sqrt{(1 - 0.05^2)}$$
$$= 63.17\,\text{rad/s}$$

$$T_d = \frac{2\pi}{\omega_d} = \frac{2\pi}{63.17} = 0.099\,\text{s}$$

$$f_d = \frac{1}{T_d} = \frac{1}{0.099} = 10.05\,\text{Hz}$$

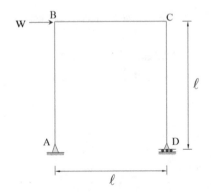

Fig. 3.18 Exercise 3.1

The required time to reach 2 mm displacement

$t = $ Number of cycles $\times T_d = 5.12 \times 0.099 = 0.506$ s.

3.4   Exercises

Exercise 3.1
Determine the natural period of the frame subjected to lateral load at point B as shown in Fig. 3.18.

Exercise 3.2
Find the displacement and velocity of the simply supported steel beam shown in Fig. 3.19 after 3 s of applying 150 mm initial displacement at the center of beam.

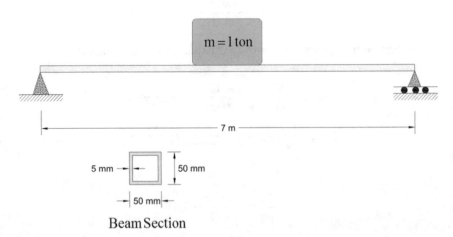

Beam Section

Fig. 3.19 Exercise 3.2

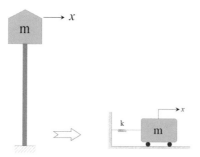

Fig. 3.20 Exercise 3.3

Exercise 3.3

The elevated tank with capacity of 5000 gallons shown in Fig. 3.20 has natural period in lateral vibration of 1.0 s when it is empty. When the tank is full of water, its period lengthens to 2.2 s. Determine the lateral stiffness of tower and the weight of tank. Neglect the mass of the supporting column.

Exercise 3.4

A mass–spring damper system consists of mass of 1 kg with unknown stiffness. The viscous damping coefficient C is struck very quickly with an impulse hammer resulting in free vibration. The displacement response is illustrated in Fig. 3.21.

(a) Determine the stiffness and damping of the system.
(b) Calculate the damping ratio using light damping approximate. Can such an approximation justify this system?

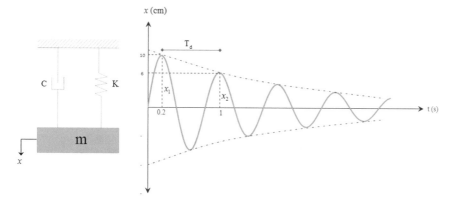

Fig. 3.21 Exercise 3.4

Chapter 4
Force Vibration of Single Degree of Freedom System

The harmonic excitation or force vibration can be described by means of sine function or by means of cosine function. The dynamic response of force vibration is defined by the following two equations of damped system:

$$m\ddot{x} + c\dot{x} + kx = F(t)$$

When force is a sine function of time:

$$m\ddot{x} + c\dot{x} + kx = F_o \sin(\omega_f t) \tag{4.1}$$

When force is a cosine function of time:

$$m\ddot{x} + c\dot{x} + kx = F_o \cos(\omega_f t) \tag{4.2}$$

F_o is the initial magnitude of external force and ω_f is the imposed frequency on the system by the force.

The total solution is the sum of complementary function and the particular integration. The complementary function will be diminished eventually because of damping. The particular integration can be performed by applying the concept of transfer function and find the required constants.

© The Editor(s) (if applicable) and The Author(s), under exclusive license
to Springer Nature Singapore Pte Ltd. 2020
F. Hejazi and T. K. Chun, *Conceptual Theories in Structural Dynamics*,
Advanced Structured Materials 135,
https://doi.org/10.1007/978-981-15-5440-7_4

4.1   Force Vibration of Undamped System

Assuming there is no friction in the structure system which means the damping of structure, c is zero in the case. Figure 4.1 shows the analytical model of undamped system subjected to force vibration.

Without considering damping, c, the Eq. (4.1) is reduced to:

$$m\ddot{x} + kx = F_o \sin(\omega_f t) \tag{4.3}$$

The solution for Eq. (4.3) is as follows:

$$x = x_{CF} + x_{PI}$$

The complementary function is as shown:

$$x_{CF} = C_1 \cos(\omega_n t) + C_2 \sin(\omega_n t) \tag{4.4}$$

Consider particular integration and by applying the concept of transfer function. The solution and its partial derivatives with respect to time, t are as shown:

$$x_{PI} = x = G \sin(\omega_f t) \tag{4.5}$$

$$\dot{x} = G \omega_f \cos(\omega_f t) \tag{4.6}$$

$$\ddot{x} = -G \omega_f^2 \sin(\omega_f t) \tag{4.7}$$

By substituting the terms (4.5) and (4.7) into Eq. (4.3) yields:

$$-mG\omega_f^2 \sin(\omega_f t) + kG \sin(\omega_f t) = F_o \sin(\omega_f t)$$

Fig. 4.1 Analytical models for undamped SDOF system under force vibration

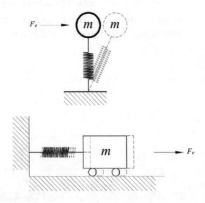

Eliminate the common term of $\sin(\omega_f t)$ from the equation above yields:

$$-m \, G \, \omega_f^2 + kG = F_o$$

Divide the terms in equation above by m:

$$-G \omega_f^2 + G\frac{k}{m} = \frac{F_o}{m}$$

By introducing $\omega_n^2 = \frac{k}{m}$, the equation above is simplified:

$$-G \omega_f^2 + G \omega_n^2 = \frac{F_o}{m}$$

Divide the terms in equation above by ω_n^2:

$$-G\frac{\omega_f^2}{\omega_n^2} + G = \frac{F_o}{m\omega_n^2}$$

By introducing $\beta = \frac{\omega_f}{\omega_n}$, the ratio of the imposed frequency to natural frequency of the system, the equation above yields:

$$-G\beta^2 + G = \frac{F_o}{m\omega_n^2}$$

From Eq. (3.13), $m\omega_n^2 = k$. By substituting this into equation above yields:

$$-G\beta^2 + G = \frac{F_o}{k}$$

Simplify the equation above yields:

$$G(1 - \beta^2) = \frac{F_o}{k}$$

Rearrange the equation above to express G in terms of other variables:

$$G = \frac{F_o}{k} \cdot \frac{1}{1 - \beta^2}$$

By substituting the equation above into Eq. (4.5) yields:

$$x_{PI} = \frac{F_o}{k} \cdot \frac{1}{1 - \beta^2} \sin(\omega_f t) \tag{4.8}$$

The total solution of undamped system subjected to force vibration [Eqs. (4.4) and (4.8)] is:

$$x = x_{CF} + x_{PI} = C_1 \cos \omega_\eta t + C_2 \sin \omega_\eta t + \frac{F_o}{k} \cdot \frac{1}{1 - \beta^2} \sin(\omega_f t) \qquad (4.9)$$

Consider the structure is at rest initially, at $t = 0$, $x = x_o = 0$
By substituting $t = 0$ and $x = 0$ into Eq. (4.9) yields:

$$0 = c_1 \cos(0) + c_2 \sin(0) + \frac{F_o}{k} \cdot \frac{1}{1 - \beta^2} \sin(0)$$

By solving the equation above yields:

$$c_1 = 0 \qquad (4.10)$$

The function of dynamic response in velocity, $\dot{x}$ can be determined by determining the partial derivative of x [Eq. (4.9)] with respect to time:

$$\dot{x} = -c_1 \omega_\eta \sin \omega_\eta t + c_2 \omega_\eta \cos \omega_\eta t + \frac{F_o}{k} \cdot \frac{\omega_f}{1 - \beta^2} \cos(\omega_f t) \qquad (4.11)$$

Under at rest condition, $\dot{x} = \dot{x}_o = 0$. By substituting $t = 0$ and $\dot{x} = 0$ into Eq. (4.11) yields:

$$0 = -c_1 \omega_\eta \sin(0) + c_2 \omega_\eta \cos(0) + \frac{F_o}{k} \cdot \frac{\omega_f}{1 - \beta^2} \cos(0)$$

By solving the equation above yields:

$$0 = c_2 \omega_\eta + \frac{F_o}{k} \cdot \frac{\omega_f}{1 - \beta^2}$$

$$c_2 = -\frac{F_o}{k \omega_\eta} \cdot \frac{\omega_f}{1 - \beta^2}$$

Rearrange the equation above yields:

$$c_2 = -\frac{F_o}{k} \cdot \frac{\omega_f}{\omega_\eta} \cdot \frac{1}{1 - \beta^2}$$

By substituting $\beta = \frac{\omega_f}{\omega_n}$ into equation above yields:

$$c_2 = -\frac{F_o}{k} \cdot \frac{\beta}{1 - \beta^2} \qquad (4.12)$$

By substituting the terms in Eqs. (4.10) and (4.12) into Eq. (4.9) yields:

$$x = -\frac{F_o}{k} \cdot \frac{\beta}{1 - \beta^2} \sin \omega_\eta t + \frac{F_o}{k} \cdot \frac{1}{1 - \beta^2} \sin(\omega_f t)$$

Rearrange the equation above yields:

$$x = \frac{F_o}{k} \cdot \frac{1}{1 - \beta^2} (\sin \omega_f t - \beta \sin \omega_\eta t) \tag{4.13}$$

From the equation above,

$\frac{F_o}{k}$ Is the static displacement, x_{static},

$\frac{1}{1-\beta^2}$ Is the dynamic magnification factor,

$\sin \omega_f t$ Is the response to imposed harmonic load,

$\beta \sin \omega_\eta t$ is the response to initial conditions.

Equation (4.13) is a function of dynamic response in terms of displacement. Therefore, Eq. (4.13) can be rewritten as:

$$x_{dynamic} = x_{static} \cdot \frac{1}{1 - \beta^2} (\sin \omega_f t - \beta \sin \omega_\eta t)$$

Rearrange the equation above yields:

$$\frac{x_{dynamic}}{x_{static}} = \frac{1}{1 - \beta^2} (\sin \omega_f t - \beta \sin \omega_\eta t)$$

where $\frac{x_{dynamic}}{x_{static}}$ is dynamic response ratio.

Now, consider the structure is not at rest initially, at $t = 0$, $x = x_o \neq 0$. By substituting $t = 0$ and $x = x_0$ into Eq. (4.9) yields:

$$x_o = c_1 \cos(0) + c_2 \sin(0) + \frac{F_o}{k} \cdot \frac{1}{1 - \beta^2} \sin(0)$$

By solving the equation above yields:

$$c_1 = x_o \tag{4.14}$$

Similarly, $\dot{x} = \dot{x}_o \neq 0$ when the structure is not at rest. By substituting $t = 0$ and $\dot{x} = \dot{x}_o$ into Eq. (4.11) yields:

$$\dot{x}_o = -x_o \omega_\eta \sin(0) + c_2 \omega_\eta \cos(0) + \frac{F_o}{k} \cdot \frac{\omega_f}{1 - \beta^2} \cos(0)$$

By solving the equation above yields:

$$\dot{x}_o = c_2 \omega_\eta + \frac{F_o}{k} \cdot \frac{\omega_f}{1 - \beta^2}$$

Rearrange the equation above yields:

$$c_2 = \frac{\dot{x}_o}{\omega_\eta} - \frac{F_o}{k} \cdot \frac{\beta}{1 - \beta^2} \tag{4.15}$$

By substituting the terms in Eqs. (4.14) and (4.15) into Eq. (4.9) yields:

$$x = x_o \cos \omega_\eta t + \left(\frac{\dot{x}_o}{\omega_\eta} - \frac{F_o}{k} \cdot \frac{\beta}{1 - \beta^2} \right) \sin \omega_\eta t + \frac{F_o}{k} \cdot \frac{1}{1 - \beta^2} \sin(\omega_f t)$$

Expand the equation above yields:

$$x = x_o \cos \omega_\eta t + \frac{\dot{x}_o}{\omega_\eta} \sin \omega_\eta t - \frac{F_o}{k} \cdot \frac{\beta}{1 - \beta^2} \sin \omega_\eta t$$
$$+ \frac{F_o}{k} \cdot \frac{1}{1 - \beta^2} \sin(\omega_f t)$$

Factorize equation above with the common term $\frac{F_o}{k} \cdot \frac{\beta}{1-\beta^2}$ yields:

$$x = x_o \cos \omega_\eta t + \frac{\dot{x}_o}{\omega_\eta} \sin \omega_\eta t$$
$$+ \frac{F_o}{k} \cdot \frac{\beta}{1 - \beta^2} \left(\sin \omega_f t - \beta \sin \omega_\eta t \right) \tag{4.16}$$

Without considering damping, c, the Eq. (4.2) is reduced to:

$$m\ddot{x} + kx = F_o \cos(\omega_f t) \tag{4.17}$$

The solution for Eq. (4.17) is as follows:

$$x = x_{CF} + x_{PI}$$

The complementary function is as shown:

$$x_{CF} = C_1 \cos \omega_\eta t + C_2 \sin \omega_\eta t \tag{4.18}$$

Consider particular integration and by applying the concept of transfer function. The solution and its partial derivatives with respect to time, t are as shown:

$$x_{PI} = x = G \cos(\omega_f t) \tag{4.19}$$

$$\dot{x} = -G\,\omega_f \sin\left(\omega_f t\right) \tag{4.20}$$

$$\ddot{x} = -G\,\omega_f^2 \cos\left(\omega_f t\right) \tag{4.21}$$

By substituting the terms Eqs. (4.19) and (4.21) into Eq. (4.17) yields:

$$-mG\omega_f^2 \cos\omega_f t + kG \cos\omega_f t = F_o \cos\omega_f t$$

Eliminate the common term $\cos\left(\omega_f t\right)$ from the equation above yields:

$$-m\,G\,\omega_f^2 + kG = F_o$$

Divide the terms in equation above by m:

$$-G\,\omega_f^2 + G\frac{k}{m} = \frac{F_o}{m}$$

By introducing $\omega_n^2 = \frac{k}{m}$, the equation above is simplified:

$$-G\,\omega_f^2 + G\,\omega_n^2 = \frac{F_o}{m}$$

Divide the terms in equation above by ω_n^2:

$$-G\frac{\omega_f^2}{\omega_n^2} + G = \frac{F_o}{m\omega_n^2}$$

By introducing $\beta = \frac{\omega_f}{\omega_n}$, the ratio of the imposed frequency to natural frequency of the system, the equation above yields:

$$-G\beta^2 + G = \frac{F_o}{m\omega_n^2}$$

From Eq. (3.13), $m\omega_n^2 = k$. By substituting this into equation above yields:

$$-G\beta^2 + G = \frac{F_o}{k}$$

Simplify the equation above yields:

$$G\left(1 - \beta^2\right) = \frac{F_o}{k}$$

Rearrange the equation above to express G in terms of other variables:

$$G = \frac{F_o}{k} \cdot \frac{1}{1 - \beta^2}$$

By substituting the equation above into Eq. (4.19) yields:

$$x_{PI} = \frac{F_o}{k} \cdot \frac{1}{1 - \beta^2} \cos \omega_f t \qquad (4.22)$$

The total solution of undamped system subjected to force vibration [Eqs. (4.4) and (4.22)] is:

$$x = x_{CF} + x_{PI} = C_1 \cos \omega_\eta t + C_2 \sin \omega_\eta t + \frac{F_o}{k} \cdot \frac{1}{1 - \beta^2} \cos \omega_f t \qquad (4.23)$$

Consider the structure is at rest initially, at $t = 0, x = x_o = 0$
By substituting $t = 0$ and $x = 0$ into Eq. (4.23) yields:

$$0 = c_1 \cos(0) + c_2 \sin(0) + \frac{F_o}{k} \cdot \frac{1}{1 - \beta^2} \cos(0)$$

By solving the equation above yields:

$$c_1 = -\frac{F_o}{k} \cdot \frac{1}{1 - \beta^2} \qquad (4.24)$$

The function of dynamic response in velocity, $\dot{x}$ can be determined by determining the partial derivative of x [Eq. (4.23)] with respect to time:

$$\dot{x} = -c_1 \omega_\eta \sin \omega_\eta t + c_2 \omega_\eta \cos \omega_\eta t - \frac{F_o}{k} \cdot \frac{\omega_f}{1 - \beta^2} \sin \omega_f t \qquad (4.25)$$

Under at rest condition, $\dot{x} = \dot{x}_o = 0$. By substituting $t = 0$ and $\dot{x} = 0$ into Eq. (4.25) yields:

$$0 = -c_1 \omega_\eta \sin(0) + c_2 \omega_\eta \cos(0) - \frac{F_o}{k} \cdot \frac{\omega_f}{1 - \beta^2} \sin(0)$$

By solving the equation above yields:

$$c_2 = 0 \qquad (4.26)$$

By substituting the terms in Eqs. (4.24) and (4.26) into Eq. (4.23) yields:

$$x = -\frac{F_o}{k} \cdot \frac{1}{1 - \beta^2} \cos \omega_\eta t + \frac{F_o}{k} \cdot \frac{1}{1 - \beta^2} \cos \omega_f t$$

Rearrange the equation above yields:

$$x = \frac{F_o}{k} \cdot \frac{1}{1 - \beta^2}\left(\cos \omega_f t - \cos \omega_\eta t\right) \tag{4.27}$$

From the equation above,

$\frac{F_o}{k}$ Is the static displacement, x_{static},

$\frac{1}{1-\beta^2}$ Is the dynamic magnification factor,

$\cos \omega_f t$ Is the response to imposed harmonic load,

$\cos \omega_\eta t$ Is the response to initial conditions.

Equation (4.27) is a function of dynamic response in terms of displacement. Therefore, Eq. (4.27) can be rewritten as:

$$x_{\text{dynamic}} = x_{\text{static}} \cdot \frac{1}{1 - \beta^2}\left(\cos \omega_f t - \cos \omega_\eta t\right)$$

Rearrange the equation above yields:

$$\frac{x_{\text{dynamic}}}{x_{\text{static}}} = \frac{1}{1 - \beta^2}\left(\cos \omega_f t - \cos \omega_\eta t\right)$$

where $\frac{x_{\text{dynamic}}}{x_{\text{static}}}$ is dynamic response ratio.

Now, consider the structure is not at rest initially, at $t = 0$, $x = x_o \neq 0$.

By substituting $t = 0$ and $x = x_0$ into Eq. (4.23) yields:

$$x_o = c_1 \cos(0) + c_2 \sin(0) + \frac{F_o}{k} \cdot \frac{1}{1 - \beta^2}\cos(0)$$

By solving the equation above yields:

$$c_1 = x_o - \frac{F_o}{k} \cdot \frac{1}{1 - \beta^2} \tag{4.28}$$

Similarly, $\dot{x} = \dot{x}_o \neq 0$ when the structure is not at rest. By substituting $t = 0$ and $\dot{x} = \dot{x}_o$ into Eq. (4.25) yields:

$$\dot{x}_o = -x_o\omega_\eta \sin(0) + c_2\omega_\eta \cos(0) - \frac{F_o}{k} \cdot \frac{\omega_f}{1 - \beta^2}\sin(0)$$

By solving the equation above yields:

$$\dot{x}_o = c_2\omega_\eta$$

Rearrange the equation above yields:

$$c_2 = \frac{\dot{x}_o}{\omega_\eta} \tag{4.29}$$

By substituting the terms in Eqs. (4.28) and (4.29) into Eq. (4.23) yields:

$$x = \left(x_o - \frac{F_o}{k} \cdot \frac{1}{1-\beta^2}\right)\cos\omega_\eta t + \frac{\dot{x}_o}{\omega_\eta}\sin\omega_\eta t + \frac{F_o}{k} \cdot \frac{1}{1-\beta^2}\cos\omega_f t$$

Expand the equation above yields:

$$x = x_o \cos\omega_\eta t - \frac{F_o}{k} \cdot \frac{1}{1-\beta^2}\cos\omega_\eta t$$
$$+ \frac{\dot{x}_o}{\omega_\eta}\sin\omega_\eta t + \frac{F_o}{k} \cdot \frac{1}{1-\beta^2}\cos\omega_f t$$

Factorize equation above with the common term $\frac{F_o}{k} \cdot \frac{1}{1-\beta^2}$ yields:

$$x = x_o \cos\omega_\eta t + \frac{\dot{x}_o}{\omega_\eta}\sin\omega_\eta t + \frac{F_o}{k} \cdot \frac{1}{1-\beta^2}(\cos\omega_f t - \cos\omega_\eta t) \tag{4.30}$$

Example 4.1 Force Vibration of Undamped System
A water tank supported on three columns subjected to harmonic load as shown in Fig. 4.2. The columns are hollow circular section with a diameter of 150 mm and thickness of 10 mm. Neglect the effect of damping. Find (a) dynamic response ratio (b) dynamic response of structure system after 10 s of applying loading.

Solution

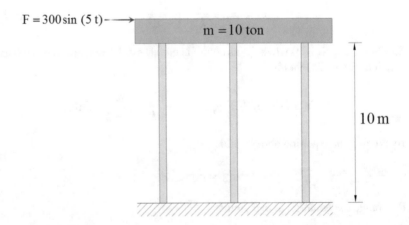

Fig. 4.2 Example 4.1

By referring to Table A.1 in Appendix, second moment of area, I for circular hollow section can be determined using the following equation:

$$I = \frac{\pi\left[D^4 - (D - 2t)^4\right]}{64} = \frac{\pi\left[0.15^4 - (0.15 - 2 \times 0.01)^4\right]}{64} = 1.083 \times 10^{-5}\,\text{m}^4$$

By referring to Table A.2 in Appendix, stiffness, k for three fixed end columns in parallel can be determined using the following equation:

$$k = 3 \times \frac{12EI}{L^3} = \frac{3 \times 12 \times 200 \times 10^9 \times 1.083 \times 10^{-5}}{10^3} = 77976\,\text{N/m}$$

The natural frequency of the structure in this case is:

$$\omega_n = \sqrt{\frac{k}{m}} = \sqrt{\frac{77976}{10000}} = 2.79\,\text{rad/s}$$

The frequency ratio is:

$$\beta = \frac{\omega_f}{\omega_n} = \frac{5}{2.79} = 1.79$$

$$\begin{aligned}
\text{DRR} &= \frac{1}{1 - \beta^2}\left(\sin \omega_f t - \beta \sin \omega_n t\right) \\
&= \frac{1}{1 - 1.79^2}\left[\sin(5 \times 10) - 1.79 \times \sin(2.79 \times 10)\right] \\
&= 0.416 < 1 \text{ ok}
\end{aligned}$$

The displacement response at 10 s is:

$$x = \frac{F_o}{k} \times \text{DRR} = \frac{300}{77,976} \times 0.416 = 1.6 \times 10^{-3}\,\text{m}$$

The velocity response at 10 s is:

$$x = \frac{F_o}{k} \cdot \frac{1}{1 - \beta^2}\left(\sin \omega_f t - \beta \sin \omega_n t\right)$$

$$\Rightarrow \dot{x} = \frac{F_o}{k} \cdot \frac{1}{1 - \beta^2}\left(\omega_f \cos \omega_f t - \beta \omega_n \cos \omega_n t\right)$$

$$\begin{aligned}
\dot{x} &= \frac{300}{77976} \cdot \frac{1}{1 - 1.79^2}\left[5 \cos(5 \times 10) - 1.79 \times 2.79 \times \cos(2.79 \times 10)\right] \\
&= -0.0165\,\text{m/s}
\end{aligned}$$

The acceleration response at 10 s is:

$$\ddot{x} = \frac{F_o}{k} \cdot \frac{1}{1-\beta^2} \left(-\omega_f^2 \sin \omega_f t + \beta \omega_\eta^2 \sin \omega_\eta t \right)$$

$$\ddot{x} = \frac{300}{77976} \cdot \frac{1}{1-1.79^2} \left[-5^2 \sin(5 \times 10) + 1.79 \times 2.79^2 \times \sin(2.79 \times 10) \right]$$

$$= -0.0203 \text{ m/s}^2$$

4.2   Force Vibration of Damped System

Structural response of a damped system subjected to external force is the basis of structural dynamic analysis and design. Figure 4.3 shows the analytical model of damped system subjected to force vibration.

The dynamic response of force vibration is defined by the Eqs. (4.1) and (4.2):

$$m\ddot{x} + c\dot{x} + kx = F_o \sin(\omega_f t)$$

$$m\ddot{x} + c\dot{x} + kx = F_o \cos(\omega_f t)$$

The solution for the sine function as shown in Eq. (4.1) is as follows:

$$x = x_{CF} + x_{PI}$$

The complementary function is as shown:

$$x_{CF} = e^{-\xi \omega_n t} \left[C_1 \cos(\omega_\eta t) + C_2 \sin(\omega_\eta t) \right] \tag{4.31}$$

Consider particular integration and by applying the concept of transfer function. The solution and its partial derivatives with respect to time, t are as shown:

Fig. 4.3 Analytical models for damped SDOF system under force vibration

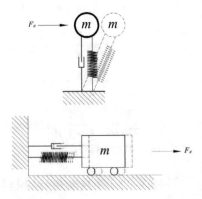

$$x_{PI} = x = G_1 \sin(\omega_f t) + G_2 \cos(\omega_f t) \tag{4.32}$$

$$\dot{x} = G_1 \omega_f \cos(\omega_f t) - G_2 \omega_f \sin(\omega_f t) \tag{4.33}$$

$$\ddot{x} = -G_1 \omega_f^2 \sin(\omega_f t) - G_2 \omega_f^2 \cos(\omega_f t) \tag{4.34}$$

By substituting the terms in Eqs. (4.32), (4.33) and (4.34) into Eq. (4.1) yields:

$$-m\,\omega_f^2 \left(G_1 \sin \omega_f t + G_2 \cos \omega_f t\right) + c\omega_f \left(G_1 \cos \omega_f t - G_2 \sin \omega_f t\right)$$
$$+ k\left(G_1 \sin \omega_f t + G_2 \cos \omega_f t\right) = F_o \sin(\omega_f t)$$

Rearrange the equation above yields:

$$\left(G_1 \sin \omega_f t + G_2 \cos \omega_f t\right)\left(k - m\omega_f^2\right) + c\omega_f \left(G_1 \cos \omega_f t - G_2 \sin \omega_f t\right) = F_o \sin(\omega_f t)$$

Expand the equation above yields:

$$G_1\left(k - m\,\omega_f^2\right)\sin \omega_f t + G_2\left(k - m\,\omega_f^2\right)\cos \omega_f t$$
$$+ G_1 c\,\omega_f \cos \omega_f t - G_2 c\,\omega_f \sin \omega_f t = F_o \sin(\omega_f t)$$

From the equation above, two equations can be obtained by comparing the terms with sine and cosine functions:

$$G_1\left(k - m\,\omega_f^2\right)\sin \omega_f t - G_2 c\,\omega_f \sin \omega_f t = F_o \sin(\omega_f t) \tag{4.35}$$

$$G_2\left(k - m\,\omega_f^2\right)\cos \omega_f t + G_1 c\,\omega_f \cos \omega_f t = 0 \tag{4.36}$$

From Eq. (4.35), eliminate the common term $\sin \omega_f t$ yields:

$$G_1\left(k - m\,\omega_f^2\right) - G_2 c\,\omega_f = F_o$$

Expand the equation above yields:

$$-m\,\omega_f^2 G_1 - G_2 c\,\omega_f + G_1 k = F_o$$

Divide the terms on both sides by m yields:

$$-\omega_f^2 G_1 - \frac{c}{m}\omega_f G_2 + \frac{k}{m}G_1 = \frac{F_o}{m}$$

By substituting Eqs. (3.14) and (3.19) into equation above yields:

$$-\omega_f^2 G_1 - 2\xi\omega_n\omega_f G_2 + \omega_n^2 G_1 = \frac{F_o}{m}$$

Divide the terms on both sides by ω_n^2 yields:

$$-\left(\frac{\omega_f}{\omega_n}\right)^2 G_1 - 2\xi\frac{\omega_f}{\omega_n}G_2 + G_1 = \frac{F_o}{\omega_n^2 m}$$

By substituting $\beta = \frac{\omega_f}{\omega_n}$ into equation above yields:

$$-\beta^2 G_1 - 2\xi\beta G_2 + G_1 = \frac{F_o}{\omega_n^2 m}$$

From Eq. (3.14), $k = \omega_n^2 m$ can be derived. By substituting this into equation above yields:

$$-\beta^2 G_1 - 2\xi\beta G_2 + G_1 = \frac{F_o}{k}$$

Simplify the equation above yields:

$$G_1\left(1 - \beta^2\right) - 2\xi\beta G_2 = \frac{F_o}{k}$$

Express G_1 in terms of other parameters yields:

$$G_1 = \frac{F_o}{k} \cdot \frac{1}{1 - \beta^2} + \frac{2\xi\beta G_2}{1 - \beta^2} \qquad (4.37)$$

From Eq. (4.36), eliminate the common term $\cos \omega_f t$ yields:

$$G_2\left(k - m\omega_f^2\right) + G_1 c\,\omega_f = 0$$

Expand the equation above yields:

$$-m\omega_f^2 G_2 + G_1 c\,\omega_f + kG_2 = 0$$

Divide the terms on both sides by m yields:

$$-\omega_f^2 G_2 + \frac{c}{m}\omega_f G_1 + \frac{k}{m}G_2 = 0$$

By substituting Eqs. (3.14) and (3.19) into equation above yields:

$$-\omega_f^2 G_2 + 2\xi\omega_n\omega_f G_1 + \omega_n^2 G_2 = 0$$

Divide the terms on both sides by ω_n^2 yields:

$$-\left(\frac{\omega_f}{\omega_n}\right)^2 G_2 + 2\xi \frac{\omega_f}{\omega_n} G_1 + G_2 = 0$$

By substituting $\beta = \frac{\omega_f}{\omega_n}$ into equation above yields:

$$-\beta^2 G_2 + 2\xi\beta G_1 + G_2 = 0$$

Simplify the equation above yields:

$$G_2\left(1 - \beta^2\right) + 2\xi\beta G_1 = 0$$

Express G_2 in terms of other parameters yields:

$$G_2 = \frac{-2\xi\beta G_1}{\left(1 - \beta^2\right)} \tag{4.38}$$

By substituting Eq. (4.38) into Eq. (4.37) yields:

$$G_1 = \frac{F_o}{k} \cdot \frac{1}{1 - \beta^2} + \frac{2\xi\beta \times \frac{-2\xi\beta G_1}{\left(1-\beta^2\right)}}{1 - \beta^2}$$

Simplify the equation above yields:

$$G_1 = \frac{F_o}{k} \cdot \frac{1}{1 - \beta^2} - \frac{(2\xi\beta)^2}{\left(1 - \beta^2\right)^2} G_1$$

Rearrange the equation above yields:

$$G_1 + \frac{(2\xi\beta)^2}{\left(1 - \beta^2\right)^2} G_1 = \frac{F_o}{k} \cdot \frac{1}{1 - \beta^2}$$

Simplify the equation above yields:

$$G_1\left[1 + \frac{(2\xi\beta)^2}{\left(1 - \beta^2\right)^2}\right] = \frac{F_o}{k} \cdot \frac{1}{1 - \beta^2}$$

$$G_1\left[\frac{\left(1 - \beta^2\right)^2 + (2\xi\beta)^2}{\left(1 - \beta^2\right)^2}\right] = \frac{F_o}{k} \cdot \frac{1}{1 - \beta^2}$$

Express G_1 in terms of other parameters yields:

$$G_1 = \frac{F_o}{k} \cdot \frac{1 - \beta^2}{\left(1 - \beta^2\right)^2 + (2\xi\beta)^2} \tag{4.39}$$

By substituting the Eq. (4.39) into Eq. (4.38) yields:

$$G_2 = \frac{-2\xi\beta}{\left(1 - \beta^2\right)} \times \frac{F_o}{k} \cdot \frac{1 - \beta^2}{\left(1 - \beta^2\right)^2 + (2\xi\beta)^2}$$

Simplify the equation above yields:

$$G_2 = -\frac{F_o}{k} \cdot \frac{2\xi\beta}{\left(1 - \beta^2\right)^2 + (2\xi\beta)^2} \tag{4.40}$$

By substituting Eqs. (4.39) and (4.40) into Eq. (4.32) yields:

$$x_{\mathrm{PI}} = \frac{F_o}{k} \cdot \frac{1 - \beta^2}{\left(1 - \beta^2\right)^2 + (2\xi\beta)^2} \sin(\omega_f t) - \frac{F_o}{k} \cdot \frac{2\xi\beta}{\left(1 - \beta^2\right)^2 + (2\xi\beta)^2} \cos(\omega_f t)$$

Simplify the equation above yields:

$$x_{\mathrm{PI}} = \frac{F_o}{k} \cdot \frac{1}{\left(1 - \beta^2\right)^2 + (2\xi\beta)^2} \left[\left(1 - \beta^2\right) \sin(\omega_f t) - 2\xi\beta \cos(\omega_f t) \right] \tag{4.41}$$

Introducing the following trigonometry relationship (Fig. 4.4):
From Fig. 4.4 above, the following relationships yield:

$$2\xi\beta = \sqrt{\left(1 - \beta^2\right)^2 + (2\xi\beta)^2} \times \sin\theta$$

$$1 - \beta^2 = \sqrt{\left(1 - \beta^2\right)^2 + (2\xi\beta)^2} \times \cos\theta$$

$$\tan\theta = \frac{2\xi\beta}{1 - \beta^2} \tag{4.42}$$

By substituting the relationships in Eq. (4.42) into Eq. (4.41) yields:

Fig. 4.4 Trigonometry relationship for $2\xi\beta$ and $1 - \beta^2$ in sine function

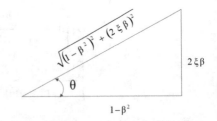

$$x_{PI} = \frac{F_o}{k} \cdot \frac{1}{\left(1 - \beta^2\right)^2 + (2\xi\beta)^2} \left[\sqrt{\left(1 - \beta^2\right)^2 + (2\xi\beta)^2} \cos\theta \sin(\omega_f t)\right.$$
$$\left. - \sqrt{\left(1 - \beta^2\right)^2 + (2\xi\beta)^2} \sin\theta \cos(\omega_f t)\right]$$

Simplify the equation above yields:

$$x_{PI} = \frac{F_o}{k} \cdot \frac{\sqrt{\left(1 - \beta^2\right)^2 + (2\xi\beta)^2}}{\left(1 - \beta^2\right)^2 + (2\xi\beta)^2} \left[\cos\theta \sin(\omega_f t) - \sin\theta \cos(\omega_f t)\right]$$

By applying the compound angle formula $\sin A \cos B \pm \cos A \sin B = \sin(A \pm B)$

$$x_{PI} = \frac{F_o}{k} \cdot \frac{1}{\sqrt{\left(1 - \beta^2\right)^2 + (2\xi\beta)^2}} \sin(\omega_f t - \theta) \tag{4.43}$$

The solution for the system subjected to external load that can be described with cosine function as shown in Eq. (4.2) is as follows:

$$x = x_{CF} + x_{PI}$$

The complementary function is as shown:

$$x_{CF} = e^{-\xi\omega_n t}\left[C_1 \cos(\omega_n t) + C_2 \sin(\omega_n t)\right] \tag{4.44}$$

Consider particular integration and by applying the concept of transfer function. The solution and its partial derivatives with respect to time, t are as shown:

$$x_{PI} = x = G_1 \cos(\omega_f t) + G_2 \sin(\omega_f t) \tag{4.45}$$

$$\dot{x} = -G_1\omega_f \sin(\omega_f t) + G_2\omega_f \cos(\omega_f t) \tag{4.46}$$

$$\ddot{x} = -G_1\omega_f^2 \cos(\omega_f t) - G_2\omega_f^2 \sin(\omega_f t) \tag{4.47}$$

By substituting the terms in Eqs. (4.45), (4.46) and (4.47) into Eq. (4.2) yields:

$$-m\omega_f^2\left(G_1 \cos\omega_f t + G_2 \sin\omega_f t\right) + c\omega_f\left(-G_1 \sin\omega_f t + G_2 \cos\omega_f t\right)$$
$$+ k\left(G_1 \cos\omega_f t + G_2 \sin\omega_f t\right) = F_o \cos(\omega_f t)$$

Rearrange the equation above yields:

$$\left(G_1 \cos\omega_f t + G_2 \sin\omega_f t\right)\left(k - m\omega_f^2\right)$$

$$+ c\omega_f\left(-G_1 \sin \omega_f t + G_2 \cos \omega_f t\right) = F_o \cos\left(\omega_f t\right)$$

Expand the equation above yields:

$$G_1\left(k - m\,\omega_f^2\right) \cos \omega_f t + G_2\left(k - m\,\omega_f^2\right) \sin \omega_f t$$
$$- G_1 c\,\omega_f \sin \omega_f t + G_2 c\,\omega_f \cos \omega_f t = F_o \cos\left(\omega_f t\right)$$

From equation above, two equations can be obtained by comparing the terms with sine and cosine functions:

$$G_1\left(k - m\,\omega_f^2\right) \cos \omega_f t + G_2 c\,\omega_f \cos \omega_f t = F_o \cos\left(\omega_f t\right) \qquad (4.48)$$

$$G_2\left(k - m\,\omega_f^2\right) \sin \omega_f t - G_1 c\,\omega_f \sin \omega_f t = 0 \qquad (4.49)$$

From Eq. (4.48), eliminate the common term $\cos \omega_f t$ yields:

$$G_1\left(k - m\,\omega_f^2\right) + G_2 c\,\omega_f = F_o$$

Expand the equation above yields:

$$-m\,\omega_f^2 G_1 + G_2 c\,\omega_f + G_1 k = F_o$$

Divide the terms on both sides by m yields:

$$-\omega_f^2 G_1 + \frac{c}{m}\omega_f G_2 + \frac{k}{m}G_1 = \frac{F_o}{m}$$

By substituting Eqs. (3.14) and (3.19) into equation above yields:

$$-\omega_f^2 G_1 + 2\xi \omega_n \omega_f G_2 + \omega_n^2 G_1 = \frac{F_o}{m}$$

Divide the terms on both sides by ω_n^2 yields:

$$-\left(\frac{\omega_f}{\omega_n}\right)^2 G_1 + 2\xi \frac{\omega_f}{\omega_n} G_2 + G_1 = \frac{F_o}{\omega_n^2 m}$$

By substituting $\beta = \frac{\omega_f}{\omega_n}$ into equation above yields:

$$-\beta^2 G_1 + 2\xi\beta G_2 + G_1 = \frac{F_o}{\omega_n^2 m}$$

From Eq. (3.14), $k = \omega_n^2 m$ can be derived. By substituting this into equation above yields:

$$-\beta^2 G_1 + 2\xi\beta G_2 + G_1 = \frac{F_o}{k}$$

Simplify the equation above yields:

$$G_1(1 - \beta^2) + 2\xi\beta G_2 = \frac{F_o}{k}$$

Express G_1 in terms of other parameters yields:

$$G_1 = \frac{F_o}{k} \cdot \frac{1}{1 - \beta^2} - \frac{2\xi\beta G_2}{1 - \beta^2} \tag{4.50}$$

From Eq. (4.49), eliminate the common term $\sin \omega_f t$ yields:

$$G_2(k - m\omega_f^2) - c\omega_f G_1 = 0$$

Expand the equation above yields:

$$-m\omega_f^2 G_2 - G_1 c\omega_f + kG_2 = 0$$

Divide the terms on both sides by m yields:

$$-\omega_f^2 G_2 - \frac{c}{m}\omega_f G_1 + \frac{k}{m}G_2 = 0$$

By substituting Eqs. (3.14) and (3.19) into equation above yields:

$$-\omega_f^2 G_2 - 2\xi\omega_n\omega_f G_1 + \omega_n^2 G_2 = 0$$

Divide the terms on both sides by ω_n^2 yields:

$$-\left(\frac{\omega_f}{\omega_n}\right)^2 G_2 - 2\xi\frac{\omega_f}{\omega_n}G_1 + G_2 = 0$$

By substituting $\beta = \frac{\omega_f}{\omega_n}$ into equation above yields:

$$-\beta^2 G_2 - 2\xi\beta G_1 + G_2 = 0$$

Simplify the equation above yields:

$$G_2(1 - \beta^2) - 2\xi\beta G_1 = 0$$

Express G_2 in terms of other parameters yields:

$$G_2 = \frac{2\xi\beta G_1}{(1-\beta^2)} \tag{4.51}$$

By substituting Eq. (4.51) into Eq. (4.50) yields:

$$G_1 = \frac{F_o}{k} \cdot \frac{1}{1-\beta^2} - \frac{2\xi\beta \times \frac{2\xi\beta G_1}{(1-\beta^2)}}{1-\beta^2}$$

Simplify the equation above yields:

$$G_1 = \frac{F_o}{k} \cdot \frac{1}{1-\beta^2} - \frac{(2\xi\beta)^2}{(1-\beta^2)^2}G_1$$

Rearrange the equation above yields:

$$G_1 + \frac{(2\xi\beta)^2}{(1-\beta^2)^2}G_1 = \frac{F_o}{k} \cdot \frac{1}{1-\beta^2}$$

Simplify the equation above yields:

$$G_1\left[1 + \frac{(2\xi\beta)^2}{(1-\beta^2)^2}\right] = \frac{F_o}{k} \cdot \frac{1}{1-\beta^2}$$

$$G_1\left[\frac{(1-\beta^2)^2 + (2\xi\beta)^2}{(1-\beta^2)^2}\right] = \frac{F_o}{k} \cdot \frac{1}{1-\beta^2}$$

Express G_1 in terms of other parameters yields:

$$G_1 = \frac{F_o}{k} \cdot \frac{1-\beta^2}{(1-\beta^2)^2 + (2\xi\beta)^2} \tag{4.52}$$

By substituting the Eq. (4.52) into Eq. (4.51) yields:

$$G_2 = \frac{2\xi\beta}{(1-\beta^2)} \times \frac{F_o}{k} \cdot \frac{1-\beta^2}{(1-\beta^2)^2 + (2\xi\beta)^2}$$

Simplify the equation above yields:

$$G_2 = \frac{F_o}{k} \cdot \frac{2\xi\beta}{(1-\beta^2)^2 + (2\xi\beta)^2} \tag{4.53}$$

By substituting Eqs. (4.52) and (4.53) into Eq. (4.45) yields:

Fig. 4.5 Trigonometry relationship for $2\xi\beta$ and $1 - \beta^2$ in cosine function

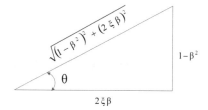

$$x_{PI} = \frac{F_o}{k} \cdot \frac{1 - \beta^2}{\left(1 - \beta^2\right)^2 + (2\xi\beta)^2} \cos(\omega_f t)$$
$$+ \frac{F_o}{k} \cdot \frac{2\xi\beta}{\left(1 - \beta^2\right)^2 + (2\xi\beta)^2} \sin(\omega_f t)$$

Simplify the equation above yields:

$$x_{PI} = \frac{F_o}{k} \cdot \frac{1}{\left(1 - \beta^2\right)^2 + (2\xi\beta)^2} \left[\left(1 - \beta^2\right) \cos(\omega_f t) + 2\xi\beta \sin(\omega_f t)\right] \qquad (4.54)$$

Introducing the following trigonometry relationship (Fig. 4.5).
From Fig. 4.5 above, the following relationships yield:

$$2\xi\beta = \sqrt{\left(1 - \beta^2\right)^2 + (2\xi\beta)^2} \times \cos\theta$$
$$1 - \beta^2 = \sqrt{\left(1 - \beta^2\right)^2 + (2\xi\beta)^2} \times \sin\theta$$
$$\tan\theta = \frac{1 - \beta^2}{2\xi\beta} \qquad (4.55)$$

By substituting the relationships in Eq. (4.55) into Eq. (4.54) yields:

$$x_{PI} = \frac{F_o}{k} \cdot \frac{1}{\left(1 - \beta^2\right)^2 + (2\xi\beta)^2}$$
$$\left[\sqrt{\left(1 - \beta^2\right)^2 + (2\xi\beta)^2} \sin\theta \cos(\omega_f t) - \cos\theta \sin(\omega_f t)\right]$$

Simplify the equation above yields:

$$x_{PI} = \frac{F_o}{k} \cdot \frac{\sqrt{\left(1 - \beta^2\right)^2 + (2\xi\beta)^2}}{\left(1 - \beta^2\right)^2 + (2\xi\beta)^2} \left[\sin\theta \cos(\omega_f t) + \cos\theta \sin(\omega_f t)\right]$$

By applying the compound angle formula $\sin A \cos B \pm \cos A \sin B = \sin(A \pm B)$

$$x_{PI} = \frac{F_o}{k} \cdot \frac{1}{\sqrt{\left(1 - \beta^2\right)^2 + (2\xi\beta)^2}} \sin\left(\omega_f t + \theta\right) \tag{4.56}$$

The total solution of damped system subjected to harmonic excitation is:

$$x = \underbrace{x_m e^{-\xi\omega_n t} \sin(\omega_d t + \varphi)}_{\text{Free Vibration (Transient)}} + \underbrace{\frac{F_o}{k} \cdot \frac{1}{\sqrt{\left(1 - \beta^2\right)^2 + (2\xi\beta)^2}} \sin\left(\omega_f t - \theta\right)}_{\text{force Vibration(Steady - State)}} \tag{4.57}$$

From Eq. (4.57), the first and second derivatives with respect to time, t of steady-state solutions are:

$$\dot{x}_{ss} = \frac{F_o}{k} \cdot \frac{\omega_f}{\sqrt{\left(1 - \beta^2\right)^2 + (2\xi\beta)^2}} \cos\left(\omega_f t\right)$$

$$\ddot{x}_{ss} = -\frac{F_o}{k} \cdot \frac{\omega_f^2}{\sqrt{\left(1 - \beta^2\right)^2 + (2\xi\beta)^2}} \sin\left(\omega_f t\right)$$

4.3 Dynamic Magnification Factor

The dynamic magnification factor for force vibration of damped system is:

$$\text{DMF} = \frac{1}{\sqrt{\left(1 - \beta^2\right)^2 + (2\xi\beta)^2}} \tag{4.58}$$

By substituting $x_{\text{static}} = \frac{F_o}{k}$ and equation above into Eqs. (4.43) and (4.56) yield:

$$x_{PI} = x_{\text{static}} \times \text{DMF} \times \sin\left(\omega_f t - \theta\right) \tag{4.59}$$

$$x_{PI} = x_{\text{static}} \times \text{DMF} \times \sin\left(\omega_f t + \theta\right) \tag{4.60}$$

Dynamic magnification factor is a function of frequency ratio and damping ratio. Generally, structure with higher damping ratio has less dynamic magnification ratio (Fig. 4.6).

Generally, there are three cases for dynamic magnification factor that will affect the dynamic response in steady-state.

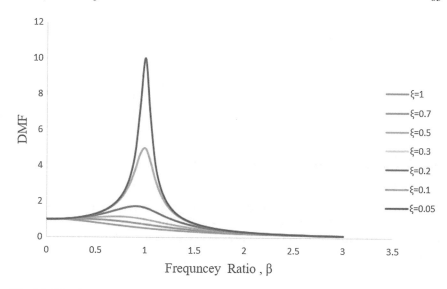

Fig. 4.6 Damping magnification factor for various frequency ratio and damping ratio

4.3.1 Frequency Ratio Nearly Equal to Zero $(\beta \cong 0)$

When the frequency of load is far less than the natural frequency of structure, slow excitation is initiated.

$$\beta = \frac{\omega_f}{\omega_n} \cong 0$$

By substituting the equation above into Eq. (4.58) yields:

$$\text{DMF} = \frac{1}{\sqrt{(1)^2 + (0)^2}} \cong 1$$

By substituting $\beta \cong 0$ into the tangent relation in (4.42) yields:

$$\tan \theta = 0$$

$$\theta = \tan^{-1} 0 = 0$$

When *DMF* is one, it means the dynamic load does not magnify the static load that is being applied to a structure. Also, when θ is zero, it means the motion and excitation are in the phase. In this case, Eq. (4.59) is reduced to:

$$x_{\text{PI}} = x_{\text{static}} \sin(\omega_f t)$$

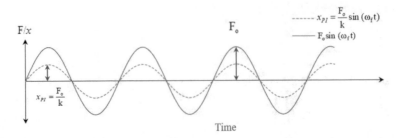

Fig. 4.7 Corresponding response for B nearly equal to 0

Figure 4.7 shows the dynamic response for the case. When the motion and excitation are in the phase, the structure is known as a stiffness control system and it will require greater stiffness to reduce the response of structure to dynamic load.

4.3.2 Frequency Ratio Is Much Greater Than One ($\beta \gg 1$)

When the frequency of load is greater than the natural frequency of structure:

$$\beta = \frac{\omega_f}{\omega_n} \gg 1$$

The damping ratio, ξ is much smaller than β. Therefore, it can be neglected. In this case, Eq. (4.58) becomes:

$$DMF = \frac{1}{\sqrt{\left(1 - \beta^2\right)^2 + (\beta)^2}}$$

Since β is very huge, $1 - \beta^2 \cong -\beta^2$. Equation above becomes:

$$DMF = \frac{1}{\sqrt{\left(-\beta^2\right)^2 + (\beta)^2}}$$

$$DMF = \frac{1}{\sqrt{\beta^4 + \beta^2}}$$

As β is very huge, $\beta^4 + \beta^2 \cong \beta^4$. Equation above becomes:

$$DMF = \frac{1}{\sqrt{\beta^4}}$$

$$\mathrm{DMF} = \frac{1}{\beta^2}$$

By substituting $\xi \cong 0$ into the tangent relation in (4.42) yields:

$$\tan \theta \cong 0$$

The value above is negative because of the denominator $1 - \beta^2 \cong -\beta^2$. Therefore, the angle is at the third quadrant and:

$$\theta \cong \pi$$

When *DMF* is one, it means the dynamic load does not magnify the static load that is being applied to a structure. Also, since θ is very near to π, it means the motion and excitation are in the opposite phase. In this case, Eq. (4.59) is transformed to:

$$x_{\mathrm{PI}} = x_{\mathrm{static}} \times \mathrm{DMF} \times \sin(\omega_f t + \pi)$$

The expression above can also be written as:

$$x_{\mathrm{PI}} = \frac{F_0}{k} \times \frac{1}{\beta^2} \times \sin(\omega_f t + \pi)$$

By substituting $\beta = \frac{\omega_f}{\omega_n}$ into equation above yields:

$$x_{\mathrm{PI}} = \frac{F_0}{k} \times \frac{1}{\left(\frac{\omega_f}{\omega_n}\right)^2} \times \sin(\omega_f t + \pi)$$

$$x_{\mathrm{PI}} = \frac{F_0}{k} \times \frac{\omega_n^2}{\omega_f^2} \times \sin(\omega_f t + \pi)$$

By substituting Eq. (3.14) into equation above yields:

$$x_{\mathrm{PI}} = \frac{F_0}{k} \times \frac{k}{m\omega_f^2} \times \sin(\omega_f t + \pi)$$

Simplify the equation above yields:

$$x_{\mathrm{PI}} = \frac{F_0}{m\omega_f^2} \times \sin(\omega_f t + \pi)$$

Figure 4.8 shows the dynamic response for the case. When the motion and excitation are in the opposite phase, the structure is known as a mass control system and it will require greater mass to reduce the response of a structure to dynamic load.

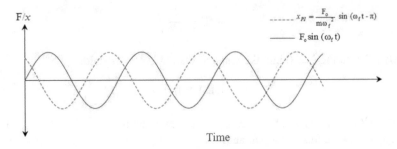

Fig. 4.8 Corresponding response for B greater than 1

4.3.3 Frequency Ratio Is Equal to One ($\beta = 1$)

When the frequency of load is equal to the natural frequency of structure:

$$\beta = \frac{\omega_f}{\omega_n} = 1$$

By substituting the equation above into Eq. (4.58) yields:

$$DMF = \frac{1}{\sqrt{(0)^2 + (2\xi)^2}}$$

$$DMF = \frac{1}{2\xi}$$

By substituting $\beta = 1$ into the tangent relation in (4.42) yields:

$$\tan\theta = \infty$$

$$\theta = \frac{\pi}{2}$$

Equation (4.59) is transformed to:

$$x_{PI} = x_{static} \times \frac{1}{2\xi} \times \sin\left(\omega_f t - \frac{\pi}{2}\right)$$

$$x_{PI} = \frac{F_o}{k} \times \frac{1}{2\xi} \times \sin\left(\omega_f t - \frac{\pi}{2}\right)$$

By substituting Eqs. (3.14) and (3.19) into equation above yields:

$$x_{PI} = \frac{F_o}{k} \times \frac{1}{2\left(\frac{C}{2\sqrt{km}}\right)} \times \sin\left(\omega_f t - \frac{\pi}{2}\right)$$

Simplify the equation above yields:

$$x_{PI} = \frac{F_o}{k} \times \frac{\sqrt{km}}{C} \times \sin\left(\omega_f t - \frac{\pi}{2}\right)$$

Rearrange the equation above yields:

$$x_{PI} = \frac{F_o}{C} \times \sqrt{\frac{m}{k}} \times \sin\left(\omega_f t - \frac{\pi}{2}\right)$$

By substituting Eq. (3.14) into equation above yields:

$$x_{PI} = \frac{F_o}{C\omega_n} \times \sin\left(\omega_f t - \frac{\pi}{2}\right)$$

When the frequency ratio is one, resonance occurs. The behaviour structure is controlled by its damping. The lower the damping ratio, the greater the magnification of load. Figure 4.9 shows the variation of the phase angle, θ according to the frequency ratio and damping of system in which the lower damping ratio record high angle phase.

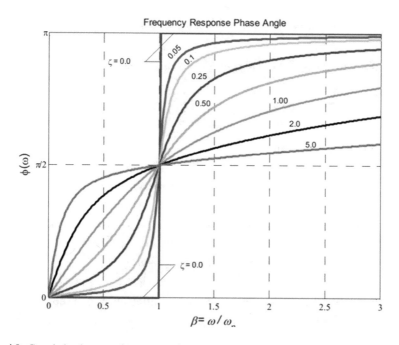

Fig. 4.9 Correlation between frequency ratio and phase angle

4.4   Dynamic Response Factors

From Eq. (4.59), let R_d be the displacement response factor and replace DMF, and express x_{static} in terms of F_0 and k yields:

$$x = x_{\text{PI}} = \frac{F_o}{k} R_d \sin(\omega_f t - \theta)$$

The expression for velocity is obtained by differentiate the equation above with respect to time, t:

$$\dot{x} = \frac{F_o}{k} \times R_d \times \omega_f \times \cos(\omega_f t - \theta)$$

By multiplying $\frac{\omega_n}{\omega_n}$ the equation above changes to:

$$\dot{x} = \frac{F_o}{k} \times R_d \times \frac{\omega_n}{\omega_n} \times \omega_f \times \cos(\omega_f t - \theta)$$

Rearrange the equation above yields:

$$\dot{x} = \frac{F_o}{k} \times R_d \times \frac{\omega_f}{\omega_n} \times \omega_n \times \cos(\omega_f t - \theta)$$

By substituting Eq. (3.14) and $\beta = \frac{\omega_f}{\omega_n}$ into equation above yields:

$$\dot{x} = \frac{F_o}{k} \times R_d \times \beta \times \sqrt{\frac{k}{m}} \times \cos(\omega_f t - \theta)$$

Simplify the equation above yields:

$$\dot{x} = \frac{F_o}{\sqrt{km}} \times R_d \times \beta \times \cos(\omega_f t - \theta)$$

By introducing velocity response factor, $R_v = R_d \times \beta$ into equation above yields:

$$\dot{x} = \frac{F_o}{\sqrt{km}} R_v \cos(\omega_f t - \theta)$$

The expression for acceleration is obtained by differentiating the equation above with respect to time, t:

$$\ddot{x} = -\frac{F_o}{\sqrt{km}} R_v \times \omega_f \times \sin(\omega_f t - \theta)$$

By multiplying $\frac{\omega_n}{\omega_n}$ the equation above changes to:

$$\ddot{x} = -\frac{F_o}{\sqrt{km}} R_v \times \omega_f \times \frac{\omega_n}{\omega_n} \times \sin(\omega_f t - \theta)$$

Rearrange the equation above yields:

$$\ddot{x} = -\frac{F_o}{\sqrt{km}} R_v \times \omega_n \times \frac{\omega_f}{\omega_n} \times \sin(\omega_f t - \theta)$$

By substituting Eq. (3.14) and $\beta = \frac{\omega_f}{\omega_n}$ into equation above yields:

$$\ddot{x} = -\frac{F_o}{\sqrt{km}} R_v \times \sqrt{\frac{k}{m}} \times \beta \times \sin(\omega_f t - \theta)$$

Simplify the equation above yields:

$$\ddot{x} = -\frac{F_o}{m} R_v \times \beta \times \sin(\omega_f t - \theta)$$

By introducing acceleration response factor, $R_a = R_v \times \beta$ into equation above yields:

$$\ddot{x} = -\frac{F_o}{m} R_a \sin(\omega_f t - \theta)$$

The relationship between acceleration, velocity and displacement response factors is:

$$R_a = R_v \times \beta = R_d \times \beta^2$$

Example 4.2 Dynamic Response Factor

Three-bay steel frame carries 10-ton load on its horizontal rigid member and subjected to harmonic load as shown in Fig. 4.10. The damping ration of frame of frame system is assumed to be 2%. Find the amplitude of vibration and the dynamic response ratio.

Solution

The stiffness of the structure is:

$$k = \frac{F_o}{x_{St}} = \frac{4000}{0.1} = 40000 \, \text{N/m}$$

The natural frequency of the structure in this case is:

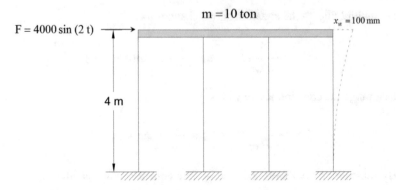

Fig. 4.10 Example 4.2

$$\omega_n = \sqrt{\frac{k}{m}} = \sqrt{\frac{40000}{10000}} = 2\,\text{rad/s}$$

The angular frequency of force is:

$$\omega_f = 2\,\text{rad/s}$$

Therefore, the frequency ratio is:

$$\beta = \frac{\omega_f}{\omega_n} = 1$$

The maximum amplitude can be determined using equation below:

$$A = x_{St} \cdot \frac{1}{\sqrt{\left(1 - \beta^2\right)^2 + (2\xi\beta)^2}} = 0.1$$

$$\times \frac{1}{\sqrt{\left(1 - 1^2\right)^2 + (2 \times 0.02 \times 1)^2}} = 2.5\,\text{m}$$

Dynamic response ratio, DRR

$$\text{DRR} = \frac{1}{\sqrt{\left(1 - \beta^2\right)^2 + (2\xi\beta)^2}}\, \sin\left(\omega_f t - \theta\right)$$

$$\frac{1}{\sqrt{\left(1 - \beta^2\right)^2 + (2\xi\beta)^2}} = \frac{1}{\sqrt{\left(1 - 1^2\right)^2 + (2 \times 0.02 \times 1)^2}} = 25$$

For $\beta \leq 1$,

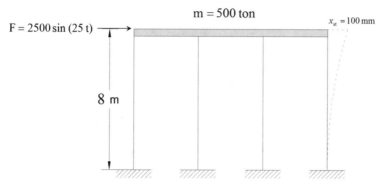

Fig. 4.11 Example 4.3

$$\theta = \tan^{-1}\left(\frac{2\xi\beta}{1-\beta^2}\right) \cong 0$$

$$\text{Dynamic response ratio} = \frac{1}{\sqrt{(1-\beta^2)^2 + (2\xi\beta)^2}}\sin(\omega_f t) = 25 \times \sin(2t)$$

Example 4.3 Force Vibration of Damped System

A steel frame shown in Fig. 4.11 carries 500-ton load on its horizontal rigid member and subjected to harmonic load. The damping ration of frame of frame system is assumed to be 2%. Determine the dimensions of columns cross-section based on the limit displacement of 30 mm. Assume the section of column is square hollow section with thickness of 5 mm.

Solution To comply with displacement limit of 30 mm, the stiffness of the structure should be more than or equal to:

$$k_{eq} = \frac{F_o}{x_{St}} = \frac{2500}{0.03} = 83333.33\,\text{N/m}$$

For a parallel system, the lateral stiffness of structure is the summation of lateral stiffness of columns. All columns are of the same type, therefore stiffness in each column can be determined using:

$$\frac{k}{4} = 20833.33\,\text{N/m}$$

By referring to Table A.2 in Appendix, stiffness for fixed end columns can be determined using the following equation:

$$k = \frac{12EI}{L^3}$$

Rearrange the equation above to solve for the second moment of area of unknown column section:

$$I = \frac{kL^3}{12E} = \frac{20833.33 \times 8^3}{12 \times 200 \times 10^9} = 4.444 \times 10^{-6} \text{m}^4$$

By referring to Table A.1 in Appendix, second moment of area, I for a rectangular hollow section can be determined using the following equation:

$$I = \frac{bh^3}{12} - \frac{(b - 2t_w)(h - 2t_f)^3}{12}$$

For square hollow section, $b = h$ and thus

$$
\begin{aligned}
I &= \frac{h^4 - (h - 2t)^4}{12} = \frac{h^4 - (h - 2 \times 0.05)^4}{12} = \frac{h^4 - (h - 0.01)^4}{12} \\
&= \frac{h^4 - (h^4 - 0.04h^3 + 6 \times 10^{-4}h^2 - 4 \times 10^{-6}h + 10^{-8})}{12} \\
&= \frac{0.04h^3 - 6 \times 10^{-4}h^2 + 4 \times 10^{-6}h - 10^{-8}}{12}
\end{aligned}
$$

Therefore,

$$4.444 \times 10^{-6} = \frac{0.04h^3 - 6 \times 10^{-4}h^2 + 4 \times 10^{-6}h - 10^{-8}}{12}$$

$$53.328 \times 10^{-6} = 0.04h^3 - 6 \times 10^{-4}h^2 + 4 \times 10^{-6}h - 10^{-8} = 0$$

$$0.04h^3 - 6 \times 10^{-4}h^2 + 4 \times 10^{-6}h - 10^{-8} - 53.328 \times 10^{-6} = 0$$

$$0.04h^3 - 6 \times 10^{-4}h^2 + 4 \times 10^{-6}h - 53.3338 \times 10^{-6} = 0$$

Simplify the equation above yields:

$$h^3 - 0.015h^2 + 10^{-4}h - 1.333 \times 10^{-3} = 0$$

Solve the cubic equation yields:

$$h = 0.115 \text{ m, adopt } h = 120 \text{ mm}$$

4.5 Exercises

Exercise 4.1
A steel frame shown in Fig. 4.12 supports 5 ton of load on its horizontal rigid member and it was also subjected to horizontal harmonic load of $F = 5000 \sin (5.3\,t)\,N$. Assuming 5% of critical damping the frame system. Determine: (a) Steady-state amplitude of vibration and (b) maximum bending stress in the column noting that the allowable stress of steel is 250 MPa.

Exercise 4.2
Three-bay steel frame carries 10-ton load on its horizontal rigid member and subjected to harmonic load as shown in Fig. 4.13. The damping ratio of frame of frame system is assumed to be 2%. Find the amplitude of vibration and the dynamic response ratio.

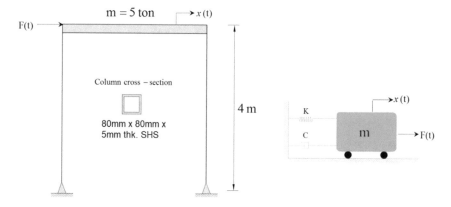

Fig. 4.12 Exercise 4.1

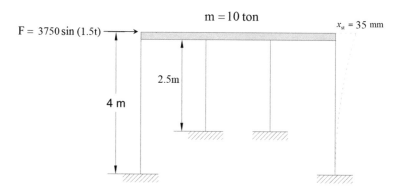

Fig. 4.13 Example 4.2

Chapter 5
Single Degree of Freedom Systems Subjected to Various Excitations

5.1 Harmonic Excitation

5.1.1 Harmonic Excitation by Generator Machine

Structural engineering does not only deal with dynamic load induced by nature, such as wind, tide and ground motion. Machine operated with a certain pattern that can be described mathematically. Unlike the natural phenomena, properties of machine vibration are consistent and can be determined easily. Therefore, dynamic load induced by machine is a type of harmonic excitation. Notable examples for this kind of machine are train, vehicle and pump. Figure 5.1 illustrates the vibration that is generated by rotating machine. The excitation by this type of machine is a function of the mass of machine, m and the mass of moving component, m'.

Speed of rotating mass is round per minutes which represents the frequency of system as shown in the following formula:

$$\omega_f = \frac{\text{rpm} \times 2\pi}{60}$$

Dynamic load is induced by the displacement of mass. In this case, it is the summation of the displacements of the entire machine, x and the moving part, x_1. Therefore,

$$x' = x_1 + x \tag{5.1}$$

From Fig. 5.1,

$$x_1 = e \sin \omega_f t$$

© The Editor(s) (if applicable) and The Author(s), under exclusive license to Springer Nature Singapore Pte Ltd. 2020
F. Hejazi and T. K. Chun, *Conceptual Theories in Structural Dynamics*, Advanced Structured Materials 135,
https://doi.org/10.1007/978-981-15-5440-7_5

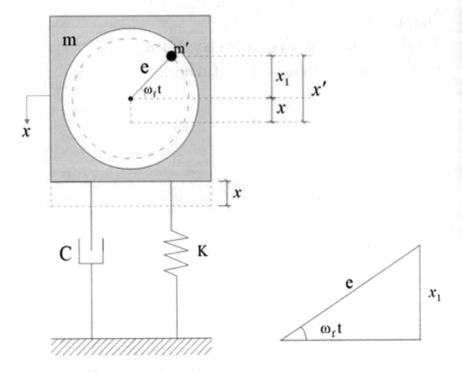

Fig. 5.1 Vibration generated by a rotating machine

By substituting the equation above into Eq. (5.1) yields:

$$x' = x + e \sin \omega_f t \tag{5.2}$$

The first and second derivatives with respect to time, t of equation above are:

$$\dot{x}' = \dot{x} + e\omega_f \cos \omega_f t \tag{5.3}$$

$$\ddot{x}' = \ddot{x} - e\omega_f^2 \sin \omega_f t \tag{5.4}$$

The general equation for motion of generating machine is:

$$(m - m')\ddot{x} + m'\ddot{x} + c\dot{x} + kx = 0$$

The term $(m - m')\ddot{x}$ is the acceleration of linear moving mass, i.e. movement of the machine along a direction, while the term $m'\ddot{x}'$ is the acceleration of rotating mass, i.e. movement of machine component about its axis of rotation.

By substituting Eqs. (5.4) into equation above yields:

$$(m - m')\ddot{x} + m'(\ddot{x} - e\omega_f^2 \sin \omega_f t) + c\dot{x} + kx = 0$$

Expand the equation above yields:

$$m\ddot{x} - m'\ddot{x} + m'\ddot{x} - m'e\omega_f^2 \sin \omega_f t + c\dot{x} + kx = 0$$

Simplify the equation above yields:

$$m\ddot{x} - m'e\omega_f^2 \sin \omega_f t + c\dot{x} + kx = 0$$

$$m\ddot{x} + c\dot{x} + kx = m'e\,\omega_f^2 \sin \omega_f t$$

By comparing the equation above with Eq. (4.1) yields:

$$F_o = m'e\,\omega_f^2$$

By substituting $x_{\text{static}} = \frac{F_o}{k}$ and relationship above into Eq. (4.59) yields:

$$x_{\text{PI}} = \frac{m'e\,\omega_f^2}{k} \times \text{DMF} \times \sin(\omega_f t - \theta)$$

In full form,

$$x = x_{\text{PI}} = \frac{m'e\,\omega_f^2}{k} \frac{1}{\sqrt{(1 - \beta^2)^2 + (2\xi\beta)^2}} \sin(\omega_f t - \theta) \tag{5.5}$$

Example 5.1 Generator Machine
Based on Fig. 5.2, a steel beam fixed on both ends supports a rotating machine. Find the displacement due to vibration of machine. Given mass of machine, $m = 500$ kg, mass of moving component, $m' = 10$ kg and it operates at 400 rpm. Take the damping ratio $\xi = 5\%$. Let $e = 150$ mm and neglect the mass of beam.

Solution
By referring to Table A.1 in Appendix, second moment of area, I for a rectangular hollow section can be determined using the following equation:

$$I = \frac{bh^3}{12} - \frac{(b - 2t_w)(h - 2t_f)^3}{12}$$
$$= \frac{0.08^4 - (0.08 - 2 \times 0.005)^4}{12} = 1.4125 \times 10^{-6}\,\text{m}^4$$

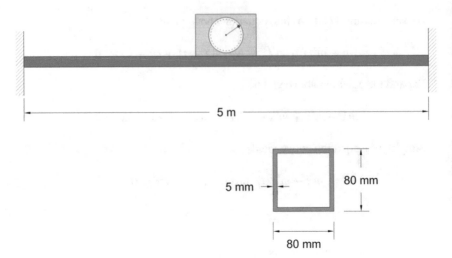

Fig. 5.2 Example 5.1

By referring to Table A.2 in Appendix, stiffness, k for fixed ends beam subjected to point load at midspan can be determined using the following equation:

$$k = \frac{192EI}{L^3}$$

$$= \frac{192 \times 200 \times 10^9 \times 1.4125 \times 10^{-6}}{5^3} = 433,920 \text{ N/m}$$

The natural frequency is:

$$\omega_n = \sqrt{\frac{k}{m}} = \sqrt{\frac{433920}{500}} = 29.46 \text{ rad/s}$$

The frequency of vibration generator in rad/s is:

$$\omega_f = \frac{400 \times 2\pi}{60} = 41.89 \text{ rad/s}$$

The frequency ratio is:

$$\beta = \frac{\omega_f}{\omega_n} = \frac{41.89}{29.46} = 1.42$$

The external force is:

$$F_o = m'e\omega_f^2 = 10 \times 0.15 \times 41.89^2 = 2632.16 \text{ N}$$

The maximum displacement is:

$$A = \frac{F_o}{k} \frac{1}{\sqrt{(1 - \beta^2)^2 + (2\xi\beta)^2}}$$
$$= \frac{2632.16}{433920} \frac{1}{\sqrt{(1 - 1.42^2)^2 + (2 \times 0.05 \times 1.42)^2}} = 5.91 \times 10^{-3}\,m$$

5.1.2 Vibration Isolation

Vibration isolation is the process whereby the vibratory effects are reduced or eliminated. There are two functions of vibration isolation as follows:

- Reduce the force magnitude transmitted from machine to the foundation.
- Reduce the magnitude of motion transmitted from foundation to a machine.

Figure 5.3a shows the case in which the source of vibration is force excitation source. The isolator such as foundation will reduce the transmitted force. On the other hand, if the source of vibration is vibrating motion of foundation (motion excitation) as shown in Fig. 5.3b. The isolator reduces the vibration of motion.

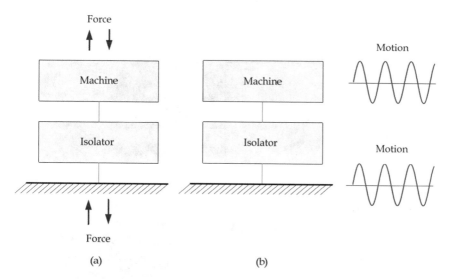

Fig. 5.3 Vibration isolation for **a** force and **b** motion excitations

5.1.3 Harmonic Motion of Foundation (Motion Excitation)

This case is when the foundation is subjected to ground time-varying motion caused by earthquakes, explosion and dynamic action of machinery. Figure 5.4 shows the analytical model of ground motion.

The harmonic motion of ground is given the following expression:

$$y = y_o \sin \omega_f t \qquad (5.6)$$

In the equation above, y_o is the maximum amplitude of ground motion and ω_f is the frequency of the foundation motion. The first derivative of equation above with respect to time, t is:

$$\dot{y} = y_o \omega_f \cos \omega_f t \qquad (5.7)$$

The structure moves with the ground. Therefore, the dynamic response of the structure is the function of relative velocity and displacement. The general equation is obtained by setting all forces shown in Fig. 5.4 equal to zero.

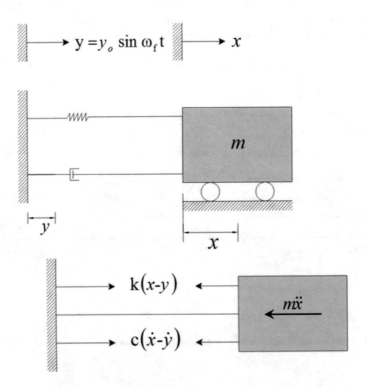

Fig. 5.4 Analytical model of ground harmonic motion

Fig. 5.5 Trigonometry relationship for $c\omega_f$ and k under motion excitation

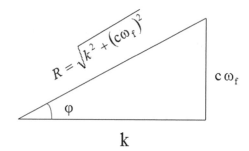

$$m\ddot{x} + c(\dot{x} - \dot{y}) + k(x - y) = 0 \tag{5.8}$$

By substituting Eqs. (5.6) and (5.7) into equation above yields:

$$m\ddot{x} + c(\dot{x} - y_o\omega_f\cos \omega_f t) + k(x - y_o\sin \omega_f t) = 0$$

Expand the equation above yields:

$$m\ddot{x} + c\dot{x} - cy_o\omega_f\cos \omega_f t + kx - ky_o\sin \omega_f t = 0$$

Rearrange the equation above yields:

$$m\ddot{x} + c\dot{x} + kx = cy_o\omega_f\cos \omega_f t + ky_o\sin \omega_f t$$

Simplify the equation above yields:

$$m\ddot{x} + c\dot{x} + kx = y_o(c\omega_f\cos \omega_f t + k\sin \omega_f t) \tag{5.9}$$

From Fig. 5.5, the following relationships yield:

$$c\omega_f = R \sin(\varphi)$$
$$k = R \cos (\varphi)$$

By substituting the relationships above into Eq. (5.9) yields:

$$m\ddot{x} + c\dot{x} + kx = y_o(R\sin \varphi \cos \omega_f t + R \cos \varphi \sin \omega_f t)$$

Simplify the equation above yields:

$$m\ddot{x} + c\dot{x} + kx = y_o R(\sin \varphi \cos \omega_f t + \cos \varphi \sin \omega_f t)$$

By applying the compound angle formula $\sin A \cos B \pm \cos A \sin B = \sin(A \pm B)$:

$$m\ddot{x} + c\dot{x} + kx = y_o R\sin(\omega_f t + \varphi)$$

By comparing the equation above with Eq., it can be seen that $y_oR\sin(\omega_f t + \varphi)$ is the function of external force. By substituting $x_{static} = \frac{F_o}{k}$ and this relationship into Eq. (4.59) yields:

$$x_{PI} = \frac{y_o R}{k} \times DMF \times \sin(\omega_f t + \varphi - \theta)$$

In full form,

$$x = x_{PI} = \frac{y_o R}{k} \frac{1}{\sqrt{(1 - \beta^2)^2 + (2\xi\beta)^2}} \sin(\omega_f t + \varphi - \theta)$$

Based on Fig. 5.5, substitute $R = \sqrt{k^2 + (c\omega_f)^2}$ into equation above yields:

$$x = \frac{y_o}{k} \frac{\sqrt{k^2 + (c\omega_f)^2}}{\sqrt{(1 - \beta^2)^2 + (2\xi\beta)^2}} \sin(\omega_f t + \varphi - \theta)$$

Simplify the equation above yields:

$$x = y_o \times \frac{\sqrt{1 + \left(\frac{c\omega_f}{k}\right)^2}}{\sqrt{(1 - \beta^2)^2 + (2\xi\beta)^2}} \sin(\omega_f t + \varphi - \theta) \tag{5.10}$$

From Eq. (3.19), the following can be expressed:

$$\frac{c\omega_f}{k} = \frac{2m\xi\omega_n\omega_f}{k}$$

By substituting Eq. (3.14) into equation above yields:

$$\frac{c\omega_f}{k} = \frac{2\xi\omega_n\omega_f}{\omega_n^2}$$

By substituting $\beta = \frac{\omega_f}{\omega_n}$ into equation above yields:

$$\frac{c\omega_f}{k} = 2\xi\beta$$

By substituting the relationship above into Eq. (5.10) yields:

$$x = y_o \times \frac{\sqrt{1 + (2\xi\beta)^2}}{\sqrt{(1 - \beta^2)^2 + (2\xi\beta)^2}} \sin(\omega_f t + \varphi - \theta)$$

Transmissibility is defined as ratio of the amplitude motion of mass x to the amplitude of ground motion, y_o .

$$T_R = \frac{x}{y_o} = \frac{\sqrt{1 + (2\xi\beta)^2}}{\sqrt{(1 - \beta^2)^2 + (2\xi\beta)^2}} \tag{5.11}$$

The influences of frequency ratio to the transmissibility are shown as follows:

$$\beta = 0 \Rightarrow T_R = 1$$

$$\beta = 1 \Rightarrow T_R = \frac{\sqrt{1 + 4\xi^2}}{2\xi} \begin{cases} \xi = 0.1 \rightarrow T_R = 5.099 \\ \xi = 0.02 \rightarrow T_R = 25.012 \\ \xi = 0.0 \rightarrow T_R = \infty \end{cases}$$

$$\beta = \sqrt{2} \Rightarrow T_R = \frac{\sqrt{1 + 8\xi^2}}{\sqrt{1 + 8\xi^2}} = 1 \tag{5.12}$$

Figure 5.6 shows the transmissibility ratio behaviour with different damping ratio in which the higher damping ratio leads to lower transmissibility as most of the vibration is dissipated through damping of the structure system.

Example 5.2 Motion Excitation
The steel frame as shown in Fig. 5.7 is subjected to sinusoidal ground motion $y(t)$ = 0.2 sin (5.3t). Determine: (a) The Transmissibility of motion to the beam girder, (b) the maximum shear force in supporting columns and (c) the maximum bending stress in columns.

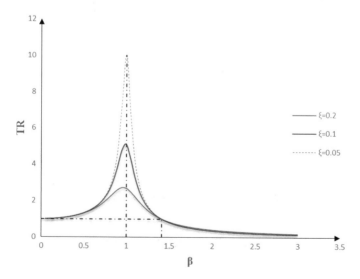

Fig. 5.6 Transmissibility ratio of structure system

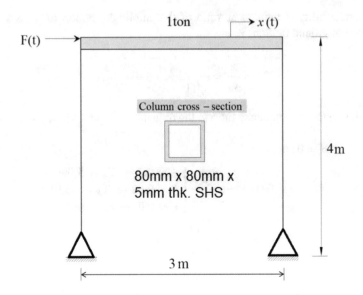

Fig. 5.7 Example 5.2

Solution

By referring to Table A.1 in Appendix, second moment of area, I for rectangular hollow section can be determined using the following equation:

$$I = \frac{bh^3}{12} - \frac{(b - 2t_w)(h - 2t_f)^3}{12}$$

$$= \frac{0.08^4 - (0.08 - 2 \times 0.005)^4}{12} = 1.4125 \times 10^{-6}\,\text{m}^4$$

By referring to Table A.2 in Appendix, stiffness for two pinned-end columns can be determined using the following equation:

$$k = 2 \times \frac{3EI}{L^3}$$

$$= 2 \times \frac{3 \times 200 \times 10^9 \times 1.4125 \times 10^{-6}}{4^3} = 26,484.37\,\text{N/m}$$

The natural frequency of the structure in this case is:

$$\omega_n = \sqrt{\frac{k}{m}} = \sqrt{\frac{26484.37}{5000}} = 2.3\,\text{rad/s}$$

The frequency ratio is:

$$\beta = \frac{\omega_f}{\omega_n} = \frac{5.3}{2.3} = 2.30$$

(a) Transmissibility

$$T_R = \frac{x}{y_o} = \frac{\sqrt{1 + (2\xi\beta)^2}}{\sqrt{(1 - \beta^2)^2 + (2\xi\beta)^2}}$$

$$= \frac{\sqrt{1 + (2 \times 0.05 \times 2.3)^2}}{\sqrt{(1 - 2.3^2)^2 + (2 \times 0.05 \times 2.3)^2}} = 0.24$$

$T_R = 0.24 < 1$, transmissibility is within isolation region.

(b) Maximum shear force in supporting columns

Maximum relative amplitude of motion is:

$$A = \frac{y_o\beta^2}{\sqrt{(1 - \beta^2)^2 + (2\xi\beta)^2}}$$

$$= \frac{0.2 \times 2.3^2}{\sqrt{(1 - 2.3^2)^2 + (2 \times 0.05 \times 2.3)^2}} = 0.24\,\text{m}$$

$$V_{\text{max}} = kA = \frac{26484.37}{2} \times 0.24 = 3178.12\,\text{N}$$

(c) Maximum bending stress in columns

For simply supported member, the bending moment can be determined using equation below:

$$M_{\text{max}} = V_{\text{max}}L = 3178.12 \times 4 = 12712.5\,\text{Nm}$$

Bending stress can be determined using equation below:

$$\sigma_{\text{max}} = \frac{M_{\text{max}}c}{I} = \frac{12712.5 \times 0.04}{1.4125 \times 10^{-6}} = 360\,\text{MPa}$$

$$\sigma_{\text{max}} = 360\,\text{MPa} > \sigma_{all} = 250\,\text{MPa}$$

Therefore, the section is not adequate under this motion excitation.

5.1.4 Force Transmissibility (Force Excitation)

The mass–spring damper system shown in Fig. 5.8 is excited by harmonic force $F(t) = F_o\sin(\omega_f t)$. The reaction Force $F_T(t)$ that is introduced to the foundation during excitation.

Let x, the function of displacement be

$$x = A\sin(\omega_f t - \theta) \tag{5.13}$$

A is relative amplitude and it is equal to $\frac{F_o}{K}\frac{1}{\sqrt{(1-\beta^2)+(2\xi\beta)^2}}$.

The first derivative of x with respect to time, t is:

$$\dot{x} = A\,\omega_f\cos(\omega_f t - \theta) \tag{5.14}$$

The reaction force F_T results from the sum of the spring force F_s and the damping force, F_d

$$F_T = F_s + F_d = kx + c\dot{x}$$

Fig. 5.8 Analytical model of force transmissibility

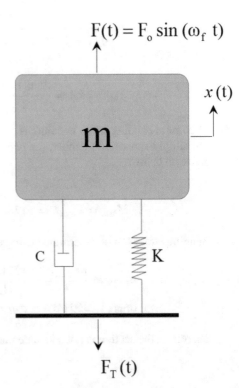

Fig. 5.9 Trigonometry relationship for $c\omega_f$ and k under force excitation

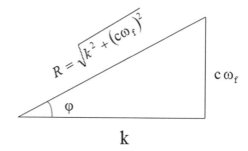

By substituting Eqs. (5.13) and (5.14) into equation above yields:

$$F_T = k A \sin(\omega_f t - \theta) + c A \omega_f \cos(\omega_f t - \theta)$$

Simplify the equation above yields:

$$F_T = A[k \sin(\omega_f t - \theta) + c\omega_f \cos(\omega_f t - \theta)] \qquad (5.15)$$

From Fig. 5.9, the following relationships yield:

$$c\omega_f = R \sin(\varphi)$$
$$k = R \cos(\varphi)$$

By substituting the relationships above into Eq. (5.15) yields:

$$F_T = A[R \cos \varphi \sin(\omega_f t - \theta) + R \sin \varphi \cos(\omega_f t - \theta)]$$

By applying the compound angle formula $\sin A \cos B \pm \cos A \sin B = \sin(A \pm B)$:

$$F_T = A R \sin(\omega_f t + \varphi - \theta)$$

By substituting $A = \frac{F_o}{K} \frac{1}{\sqrt{(1-\beta^2)+(2\xi\beta)^2}}$ into equation above yields:

$$F_T = F_o \frac{\sqrt{1 + (2\xi\beta)^2}}{\sqrt{(1 - \beta^2) + (2\xi\beta)^2}} \sin(\omega_f t + \varphi - \theta)$$

The force transmissibility ratio is defined as below:

$$\frac{F_T}{F_o} = \frac{\sqrt{1 + (2\xi\beta)^2}}{\sqrt{(1 - \beta^2) + (2\xi\beta)^2}}$$

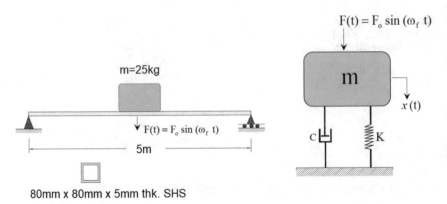

m=25kg

F(t) = F$_o$ sin (ω_f t)

5m

80mm x 80mm x 5mm thk. SHS

Fig. 5.10 Example 5.3

The effect of frequency ratio is as shown in (5.12).

Example 5.3 Force Excitation
A machine with mass of 25 kg is mounted on simply supported beam as shown in
Fig. 5.10. A piston moves up and down in the machine produces a harmonic force
of 250 N at frequency of $\omega_f = 60$ rad/s. Neglect the weight of beam and assuming
10% of the critical damping. Determine: (a) the amplitude of motion of machine, (b)
the force transmitted to the beam supports.

Solution
By referring to Table A.1 in Appendix, second moment of area, I for rectangular
hollow section can be determined using the following equation:

$$I = \frac{bh^3}{12} - \frac{(b - 2t_w)(h - 2t_f)^3}{12}$$

$$= \frac{0.08^4 - (0.08 - 2 \times 0.005)^4}{12} = 1.4125 \times 10^{-6} \, \text{m}^4$$

By referring to Table A.2 in Appendix, bending stiffness for simply supported
beam subjected to point load at midspan can be determined using the following
equation:

$$k = \frac{48EI}{L^3}$$

$$= \frac{48 \times 200 \times 10^9 \times 1.4125 \times 10^{-6}}{5^3} = 108,480 \, \text{N/m}$$

The natural frequency of the structure in this case is:

$$\omega_n = \sqrt{\frac{k}{m}} = \sqrt{\frac{108,480}{25}} = 65.87 \, \text{rad/s}$$

The frequency ratio is:

$$\beta = \frac{\omega_f}{\omega_n} = \frac{5.3}{60} = 0.08$$

Force transmissibility ratio is:

$$\begin{aligned}
T_R = \frac{F_T}{F_o} &= \frac{\sqrt{1 + (2\xi\beta)^2}}{\sqrt{(1 - \beta^2)^2 + (2\xi\beta)^2}} \\
&= \frac{\sqrt{1 + (2 \times 0.1 \times 0.08)^2}}{\sqrt{(1 - 0.08^2)^2 + (2 \times 0.1 \times 0.08)^2}} = 1
\end{aligned}$$

$T_R = 1$, the excited force is fully transmitted to the supports

(a) Maximum relative amplitude of motion is:

$$\begin{aligned}
A &= \frac{F_0}{k} \frac{1}{\sqrt{(1 - \beta^2)^2 + (2\xi\beta)^2}} \\
&= \frac{250}{108,480} \frac{1}{\sqrt{(1 - 0.08^2)^2 + (2 \times 0.1 \times 0.08)^2}} \\
&= 2.31 \times 10^{-3} \, \text{m} = 2.31 \, \text{mm}
\end{aligned}$$

Phase angle is:

$$\begin{aligned}
\theta &= \tan^{-1}\left(\frac{2\xi\beta}{1 - \beta^2}\right) \\
&= \tan^{-1}\left(\frac{2 \times 0.1 \times 0.08}{1 - 0.08^2}\right) = 0.92°
\end{aligned}$$

(b) Transmitted force is:

$$F_T = F_0 T_R = 25 \, \text{kg} \times 9.81 \, \text{N/kg} \times 1 = 245.25 \, \text{N}$$

5.1.5   Vibration Measurement Instrument

Experimentally, the measurement of the motion or displacement can be performed by use of LVDTS or other techniques such as measurement by leaser. The concept of vibration measurement is illustrated in Fig. 5.11.

The general equation of motion excitation is as defined in Eq. (5.8). By subtracting the inertial force generated by external motion of y, $m\ddot{y}$ on both sides yields:

$$m\ddot{x} + c(\dot{x} - \dot{y}) + k(x - y) - m\ddot{y} = -m\ddot{y}$$

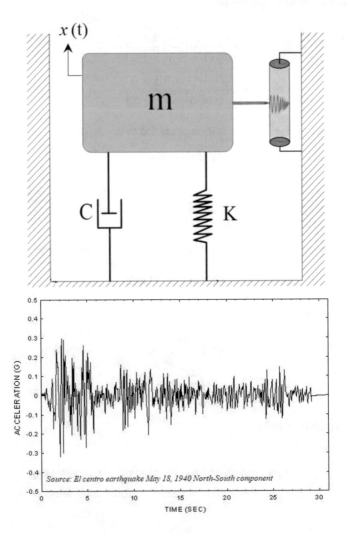

Fig. 5.11 Analytical model for vibration measurement

Simplify the equation above yields:

$$m(\ddot{x} - \ddot{y}) + c(\dot{x} - \dot{y}) + k(x - y) = -m\ddot{y}$$

Let $(\ddot{x} - \ddot{y}) = \ddot{z}, (\dot{x} - \dot{y}) = \dot{z}, (x - y) = z$, and by substituting these relationships into equation above yields:

$$m\ddot{z} + c\dot{z} + kz = -m\ddot{y} \tag{5.16}$$

The derivative of Eq. (5.7) with respect to time, t is the function of acceleration for y:

$$\ddot{y} = -y_o\omega_f^2 \sin \omega_f t \tag{5.17}$$

By substituting Eq. (5.17) into Eq. (5.16) yields:

$$m\ddot{z} + c\dot{z} + kz = my_o\omega_f^2 \sin \omega_f t$$

In a similar form as x_{PI} in Eq. (5.5), the solution of z, z_{PI} is expressed as follows:

$$z_{PI} = \frac{m\omega_f^2 y_o}{k} \frac{1}{\sqrt{(1 - \beta^2)^2 + (2\xi\beta)^2}} \sin\left(\omega_f t - \theta\right)$$

By substituting Eq. (3.14) into equation above yields:

$$z_{PI} = \frac{y_o\omega_f^2}{\omega_n^2} \frac{1}{\sqrt{(1 - \beta^2)^2 + (2\xi\beta)^2}} \sin\left(\omega_f t - \theta\right)$$

By substituting $\beta = \frac{\omega_f}{\omega_n}$ into equation above yields:

$$z_{PI} = y_o \frac{\beta^2}{\sqrt{(1 - \beta^2)^2 + (2\xi\beta)^2}} \sin\left(\omega_f t - \theta\right) \tag{5.18}$$

The dynamic response ratio is defined as follows:

$$\frac{z_{PI}}{y_o} = \frac{\beta^2}{\sqrt{(1 - \beta^2)^2 + (2\xi\beta)^2}} \sin\left(\omega_f t - \theta\right)$$

By comparing Eq. (5.18) above with Eq. (5.13) yields:

$$A = \frac{y_o \cdot \beta^2}{\sqrt{(1 - \beta^2)^2 + (2\xi\beta)^2}}$$

5.2   Impact Loads

5.2.1   Duhamel's Integral for Undamped System

Impulsive loading is a kind of load that is applied in short duration of time. The corresponding impulse of this type of load is defended as product of force and time of its duration. Figure 5.12 shows the general function of impulsive load.

By Newton's second law of motion,

$$F = ma$$

Impulsive load is a function of time, τ. Therefore, the equation above can be expressed as

$$F(\tau) = m\frac{dv}{d\tau}$$

The incremental velocity, dv is thus expressed as:

$$dv = \frac{F(\tau)d\tau}{m} \tag{5.19}$$

For undamped system under free vibration, displacement x is expressed as Eq. (3.10). Initial displacement, x_0 is zero, and thus the equation becomes:

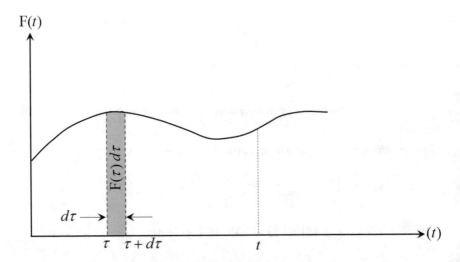

Fig. 5.12 General load function as impulsive loading

$$x = \frac{\dot{x}_0}{\omega_\eta} \sin \omega_\eta t$$

From equation above

$$dx = \frac{d\dot{x}_o}{\omega_n} \sin \omega_n (t - \tau)$$

For a general load function varies with time, displacement over certain period of time can be determined by calculating the area under the function curve. Mathematically, it can be expressed in integral form as follows:

$$x = \int_0^t \frac{d\dot{x}_o}{\omega_n} \sin \omega_n (t - \tau) d\tau$$

By substituting Eq. (5.19) into equation above yields:

$$x = \int_0^t \frac{F(\tau)}{m\omega_n} \sin \omega_n (t - \tau) d\tau$$

Simplify the equation above and yields the Duhamel's integral for undamped system:

$$x = \frac{1}{m\omega_n} \int_0^t F(\tau) \sin \omega_n (t - \tau) d\tau \qquad (5.20)$$

5.2.2 Constant Load

By considering the case of a constant force with magnitude F_o applied suddenly to undamped mass–spring system at time $t = 0$ as shown in Fig. 5.13.

The function of constant load is:

$$F(\tau) = F_o$$

Therefore, the Duhamel's integral (Eq. 5.20) for this load case is:

$$x = \frac{1}{m\omega_n} \int_0^t F_o \sin \omega_n (t - \tau) d\tau$$

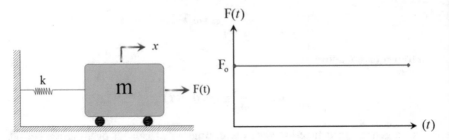

Fig. 5.13 Mass–spring system acted upon by constant load

Rearrange the equation above yields:

$$x = \frac{F_o}{m\omega_n} \int\limits_0^t \sin \omega_n (t - \tau) d\tau$$

By performing integration for the equation above yields:

$$x = \frac{F_o}{m\omega_n} \times -\frac{1}{\omega_n} \times -[\cos \omega_n (t - \tau)]_0^t$$

Simplify the equation above yields:

$$x = \frac{F_o}{m\omega_n^2} \times [\cos \omega_n (t - \tau)]_0^t$$

Solve the equation above yields:

$$x = \frac{F_o}{m\omega_n^2} \times (1 - \cos \omega_n t)$$

By substituting Eq. (3.14) into equation above yields:

$$x = \frac{F_o}{k} \times (1 - \cos \omega_n t)$$

By substituting $x_{\text{static}} = \frac{F_0}{k}$ into equation above yields:

$$x = x_{\text{static}} \times (1 - \cos \omega_n t)$$

By substituting Eq. (3.15) into equation above yields:

$$x = x_{\text{static}} \times \left(1 - \cos 2\pi \frac{t}{T}\right)$$

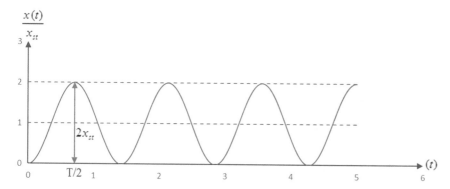

Fig. 5.14 Dynamic response of undamped system due to sudden applied constant force

The DMF for this load case is:

$$\text{DMF} = \left(1 - \cos 2\pi \frac{t}{T}\right)$$

The influence of time, t to the dynamic responses in term of displacement for this load case are as follows:

$$t = 0 \Rightarrow x = x_{st} \times \left(1 - \cos 2\pi \frac{0}{T}\right) \Rightarrow x = 0$$

$$t = T \Rightarrow x = x_{st} \times \left(1 - \cos 2\pi \frac{T}{T}\right) \Rightarrow x = 0$$

$$t = \frac{T}{2} \Rightarrow x = x_{st} \times \left(1 - \cos 2\pi \frac{T}{2T}\right) \Rightarrow x = 2x_{st}$$

Figure 5.14 shows the dynamic response of undamped system subjected to suddenly apply constant force.

5.2.3 Rectangular Load

In this case, the constant load F_o suddenly applied on undamped system but only during a limited time duration, t_d as shown in Fig. 5.15.
The function of rectangular load is:

$$F(\tau) = F_o, 0 \leq t \leq t_d$$

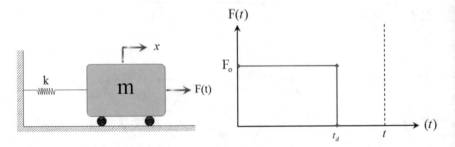

Fig. 5.15 Rectangular constant force acting on mass–spring system

This load case is divided into two phases: the first phase is for timeframe $0 \le t \le t_d$, where the load applied is equal to F_o. The second phase is for timeframe $t > t_d$, where no load is applied.

For the first phase, the Duhamel's integral (Eq. (5.20)) for this load case is:

$$x = \frac{1}{m\omega_n} \int_0^t F_o \sin \omega_n (t - \tau) d\tau$$

Rearrange the equation above yields:

$$x = \frac{F_o}{m\omega_n} \int_0^t \sin \omega_n (t - \tau) d\tau$$

By performing integration for the equation above yields:

$$x = \frac{F_o}{m\omega_n} \times -\frac{1}{\omega_n} \times -[\cos \omega_n (t - \tau)]_0^t$$

Simplify the equation above yields:

$$x = \frac{F_o}{m\omega_n^2} \times [\cos \omega_n (t - \tau)]_0^t$$

Solve the equation above yields:

$$x = \frac{F_o}{m\omega_n^2} \times (1 - \cos \omega_n t)$$

By substituting Eq. (3.14) into equation above yields:

$$x = \frac{F_o}{k} \times (1 - \cos \omega_n t)$$

For the second phase, the effect of the first phase is the initial condition for the second phase. By substituting $t = t_d$ into equation above yields the initial displacement of the second phase:

$$x_o = \frac{F_o}{k} \times (1 - \cos \omega_n t_d) \tag{5.21}$$

The derivative of equation above with respect to time, t is:

$$\dot{x}_o = \frac{F_o}{k} \times \omega_n \sin \omega_n t_d \tag{5.22}$$

For undamped system under free vibration as expressed in Eq. (3.10), the equation for the timeframe $t - t_d$ yields:

$$x = x_o \cos \omega_n (t - t_d) + \frac{\dot{x}_o}{\omega_n} \sin \omega_n (t - t_d)$$

By substituting Eqs. (5.21) and (5.22) into equation above yields:

$$x = \frac{F_o}{k} \times (1 - \cos \omega_n t_d) \cos \omega_n (t - t_d)$$
$$+ \frac{F_o}{k} \times \sin \omega_n t_d \sin \omega_n (t - t_d)$$

Simplify the equation above yields:

$$x = \frac{F_o}{k} [(1 - \cos \omega_n t_d) \cos \omega_n (t - t_d) + \sin \omega_n t_d \sin \omega_n (t - t_d)]$$

By expanding the equation above yields:

$$x = \frac{F_o}{k} [\cos \omega_n (t - t_d) - \cos \omega_n (t - t_d) \cos \omega_n t_d$$
$$+ \sin \omega_n t_d \sin \omega_n (t - t_d)]$$

Rearrange the equation above yields:

$$x = \frac{F_o}{k} \{\cos \omega_n (t - t_d)$$
$$- [\cos \omega_n (t - t_d) \cos \omega_n t_d - \sin \omega_n t_d \sin \omega_n (t - t_d)]\}$$

By applying the compound angle formula $\cos(A \pm B) = \cos A \cos B \mp \sin A \sin B$ to the equation above yields:

$$x = \frac{F_o}{k}[\cos \omega_n (t - t_d) - \cos \omega_n t]$$

By substituting Eq. (3.15) into equation above yields:

$$x = \frac{F_o}{k}\left[\cos 2\pi \left(\frac{t}{T} - \frac{t_d}{T}\right) - \cos 2\pi \frac{t}{T}\right]$$

By substituting $x_{\text{static}} = \frac{F_o}{k}$ into equation above yields:

$$x = x_{st} \times \left[\cos 2\pi \left(\frac{t}{T} - \frac{t_d}{T}\right) - \cos 2\pi \frac{t}{T}\right]$$

The DMF for the load case is:

$$\text{DMF} = \left[\cos 2\pi \left(\frac{t}{T} - \frac{t_d}{T}\right) - \cos 2\pi \frac{t}{T}\right]$$

5.2.4   Triangular Load

The third case is when the undamped system is subjected to an initial value of F_o and that decrease linearly till it becomes zero value at time t_d as shown in Fig. 5.16. The function of rectangular load is:

$$F(\tau) = F_o\left(1 - \frac{\tau}{t_d}\right), 0 \le t \le t_d$$

This load case is divided into two phases: the first phase is for timeframe $0 \le t \le t_d$, where the load applied is equal to F_o. The second phase is for timeframe $t > t_d$, where no load is applied.

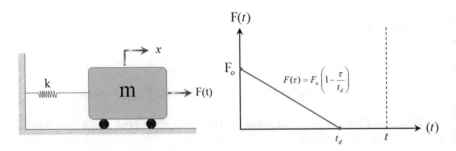

Fig. 5.16 Triangular dynamic force acted upon undamped oscillator

For the first phase, the Duhamel's integral (Eq. (5.20)) for this load case is:

$$x = \frac{1}{m\omega_n} \int_0^t F_o\left(1 - \frac{\tau}{t_d}\right) \sin \omega_n(t - \tau) d\tau$$

Simplify the equation above yields:

$$x = \frac{F_o}{m\omega_n} \int_0^t \left(1 - \frac{\tau}{t_d}\right) \sin \omega_n(t - \tau) d\tau$$

Expand the equation above yields:

$$x = \frac{F_o}{m\omega_n} \int_0^t \sin \omega_n(t - \tau) - \frac{\tau}{t_d} \sin \omega_n(t - \tau) d\tau$$

Rearrange the equation above yields:

$$x = \frac{F_o}{m\omega_n} \left[\int_0^t \sin \omega_n(t - \tau) d\tau - \int_0^t \frac{\tau}{t_d} \sin \omega_n(t - \tau) d\tau \right]$$

$$x = \frac{F_o}{m\omega_n} \int_0^t \sin \omega_n(t - \tau) d\tau - \frac{F_o}{m\omega_n t_d} \int_0^t \tau \sin \omega_n(t - \tau) d\tau \qquad (5.23)$$

The product of integration for the first part of the equation above is:

$$\frac{F_o}{m\omega_n} \int_0^t \sin \omega_n(t - \tau) d\tau = \frac{F_o}{k}(1 - \cos \omega_n t). \qquad (5.24)$$

The product of integration for the second part of the Eq. (5.23) can be determined with integration by parts $\int u dv = uv - \int v du$:

Let $u = \tau$ and $dv = \sin \omega_n(t - \tau)$, the following terms yielded:

$$du = d\tau$$

$$v = \frac{1}{\omega_n} \cos \omega_n(t - \tau)$$

In general form of $\int u\,dv = uv - \int v\,du$:

$$-\frac{F_o}{m\omega_n t_d}\int_0^t \tau \sin\omega_n(t-\tau)d\tau = -\frac{F_o}{m\omega_n t_d}\left[\frac{\tau}{\omega_n}\cos\omega_n(t-\tau)\right.$$

$$\left. -\int \frac{1}{\omega_n}\cos\omega_n(t-\tau)d\tau\right]$$

$$-\frac{F_o}{m\omega_n t_d}\int_0^t \tau \sin\omega_n(t-\tau)d\tau = -\frac{F_o}{m\omega_n t_d}\left[\frac{\tau}{\omega_n}\cos\omega_n(t-\tau)\right.$$

$$\left. -\frac{1}{\omega_n}\int \cos\omega_n(t-\tau)d\tau\right]$$

By performing integration accordingly:

$$-\frac{F_o}{m\omega_n t_d}\int_0^t \tau \sin\omega_n(t-\tau)d\tau = -\frac{F_o}{m\omega_n t_d}\left[\frac{\tau}{\omega_n}\cos\omega_n(t-\tau)+\frac{1}{\omega_n^2}\sin\omega_n(t-\tau)\right]_0^t$$

Solve the equation above yields:

$$-\frac{F_o}{m\omega_n t_d}\int_0^t \tau \sin\omega_n(t-\tau)d\tau = -\frac{F_o}{m\omega_n t_d}\left[\frac{\tau}{\omega_n}-\frac{1}{\omega_n^2}\sin\omega_n t\right]$$

By factorizing the equation above with the term $-\frac{1}{\omega_n}$ yields:

$$-\frac{F_o}{m\omega_n t_d}\int_0^t \tau\sin\omega_n(t-\tau)d\tau = \frac{F_o}{m\,\omega_n t_d}\times\frac{1}{\omega_n}\left(\frac{\sin\omega_n t}{\omega_n}-\tau\right)$$

$$-\frac{F_o}{m\omega_n t_d}\int_0^t \tau\sin\omega_n(t-\tau)d\tau = \frac{F_o}{m\,\omega_n^2 t_d}\left(\frac{\sin\omega_n t}{\omega_n}-\tau\right)$$

By substituting Eq. (3.14) into equation above yields:

$$-\frac{F_o}{m\omega_n t_d}\int_0^t \tau\sin\omega_n(t-\tau)d\tau = \frac{F_o}{k t_d}\left(\frac{\sin\omega_n t}{\omega_n}-\tau\right) \qquad (5.25)$$

By substituting Eqs. (5.24) and (5.25) into Eq. (5.23) yields:

$$x = \frac{F_o}{k}(1 - \cos \omega_n t) + \frac{F_o}{kt_d}\left(\frac{\sin \omega_n t}{\omega_n} - \tau\right)$$

Simplify the equation above yields:

$$x = \frac{F_o}{k}\left[(1 - \cos \omega_n t) - \frac{1}{t_d}\left(\frac{\sin \omega_n t}{\omega_n} - \tau\right)\right] \tag{5.26}$$

For the second phase, the effect of the first phase is the initial condition for the second phase. By substituting $t = t_d$ into equation above yields the initial displacement of the second phase:

$$x_o = \frac{F_o}{k}(1 - \cos \omega_n t_d) + \frac{F_o}{kt_d}\left(\frac{\sin \omega_n t_d}{\omega_n} - t_d\right) \tag{5.27}$$

The derivative of Eq. (5.26) with respect to time, t is:

$$\dot{x}_o = \frac{F_o}{k} \times \omega_n \sin \omega_n t + \frac{F_o}{kt_d}(\cos \omega_n t - 1)$$

The initial velocity for the second phase can be obtained by substituting $t = t_d$ into equation above:

$$\dot{x}_o = \frac{F_o}{k} \times \omega_n \sin \omega_n t_d + \frac{F_o}{kt_d}(\cos \omega_n t_d - 1) \tag{5.28}$$

For undamped system under free vibration as expressed in Eq. (3.10), the equation for the timeframe $t - t_d$ yields:

$$x = x_o \cos \omega_n(t - t_d) + \frac{\dot{x}_o}{\omega_n}\sin \omega_n(t - t_d)$$

By substituting Eqs. (5.27) and (5.28) into equation above yields:

$$x = \left[\frac{F_o}{k}(1 - \cos \omega_n t_d) + \frac{F_o}{kt_d}\left(\frac{\sin \omega_n t_d}{\omega_n} - t_d\right)\right]\cos \omega_n(t - t_d)$$
$$+ \frac{1}{\omega_n}\left[\frac{F_o}{k} \times \omega_n \sin \omega_n t_d + \frac{F_o}{kt_d}(\cos \omega_n t_d - 1)\right]\sin \omega_n(t - t_d)$$

By expanding the equation above yields:

$$x = \frac{F_o}{k}\cos \omega_n(t - t_d) - \frac{F_o}{k}\cos \omega_n(t - t_d)\cos \omega_n t_d$$

$$+ \frac{F_o}{k\omega_n t_d} \cos \omega_n (t - t_d) \sin \omega_n t_d - \frac{F_o}{k} \cos \omega_n (t - t_d)$$

$$+ \frac{F_o}{k} \sin \omega_n (t - t_d) \sin \omega_n t_d + \frac{F_o}{k\omega_n t_d} \sin \omega_n (t - t_d) \cos \omega_n t_d$$

$$- \frac{F_o}{k\omega_n t_d} \sin \omega_n (t - t_d)$$

Rearrange the equation above yields:

$$x = -\frac{F_o}{k} [\cos \omega_n (t - t_d) \cos \omega_n t_d - \sin \omega_n (t - t_d) \sin \omega_n t_d]$$

$$+ \frac{F_o}{k\omega_n t_d} [\cos \omega_n (t - t_d) \sin \omega_n t_d + \sin \omega_n (t - t_d) \cos \omega_n t_d]$$

$$- \frac{F_o}{k\omega_n t_d} \sin \omega_n (t - t_d)$$

By applying the compound angle formulas $\sin A \cos B \pm \cos A \sin B = \sin(A \pm B)$ and $\cos(A \pm B) = \cos A \cos B \mp \sin A \sin B$:

$$x = -\frac{F_o}{k} [\cos(\omega_n t - \omega_n t_d + \omega_n t_d)]$$

$$+ \frac{F_o}{k\omega_n t_d} [\sin(\omega_n t - \omega_n t_d + \omega_n t_d)] - \frac{F_o}{k\omega_n t_d} \sin \omega_n (t - t_d)$$

Simplify the equation above yields:

$$x = -\frac{F_o}{k} \cos(\omega_n t) + \frac{F_o}{k\omega_n t_d} \sin(\omega_n t) - \frac{F_o}{k\omega_n t_d} \sin \omega_n (t - t_d)$$

Simplify the equation above by grouping the coefficients of sin and cosine functions yields:

$$x = \frac{F_o}{k\omega_n t_d} [\sin(\omega_n t) - \sin \omega_n (t - t_d)] - \frac{F_o}{k} \cos(\omega_n t)$$

Factorize the equation above with the term $\frac{F_o}{k}$ yields:

$$x = \frac{F_o}{k} \left[\frac{1}{\omega_n t_d} [\sin(\omega_n t) - \sin \omega_n (t - t_d)] - \cos(\omega_n t) \right]$$

By substituting $x_{static} = \frac{F_o}{k}$ into equation above yields:

$$x = x_{static} \left[\frac{1}{\omega_n t_d} [\sin(\omega_n t) - \sin \omega_n (t - t_d)] - \cos(\omega_n t) \right]$$

The DMF for the load case is:

$$\mathrm{DMF} = \frac{1}{\omega_n t_d}[\sin(\omega_n t) - \sin \omega_n (t - t_d)] - \cos(\omega_n t)$$

Figure 5.17 gives the chart of maximum response for single degree of freedom system under rectangular and triangular loading. This chart is called response spectra chart. This chart has been developed based on the previous mathematical approach for the impulsive loads for short duration of time. For a short time loading, the effect of damping is neglected. The maximum dynamic load factor reflects the first peak of response and normally the magnitude of damping in the structure system is not enough to decrease this value.

Example 5.4 Impact Load

A steel signboard has load 1 ton and supported on 10 m column height as shown in Fig. 5.18. The signboard was subjected to 100 kN rectangular impulsive load caused

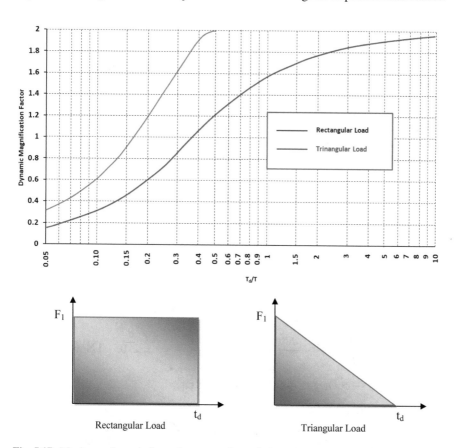

Fig. 5.17 Maximum dynamic factor for rectangular and triangular forces (Paz and Leigh 2004)

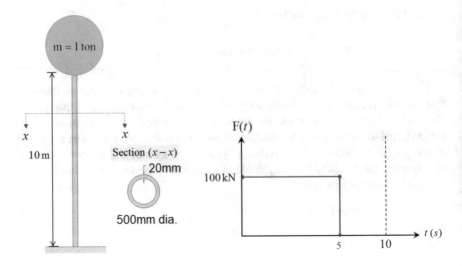

Fig. 5.18 Example 5.4

by a vehicle. Neglect the weight of column. Determine the dynamic response of the signboard after it has been hit.

Solution

By referring to Table A.1 in Appendix, second moment of area, I for circular hollow section can be determined using the following equation:

$$I = \frac{\pi d^4}{64} - \frac{\pi (d - 2t)^4}{64}$$

$$= \frac{\pi (0.5)^4}{64} - \frac{\pi (0.5 - 2 \times 0.02)^4}{64}$$

$$= 8.7 \times 10^{-7} \, \text{m}^4$$

By referring to Table A.2 in Appendix, stiffness, k for cantilever with point load at free end can be determined using the following equation:

$$k = \frac{3EI}{L^3} = \frac{3 \times 200 \times 10^9 \times 8.7 \times 10^{-4}}{10^3} = 522,057.3 \, \text{N/m}$$

The natural frequency of the structure in this case is:

$$\omega_n = \sqrt{\frac{k}{m}} = \sqrt{\frac{522057.3}{10000}} = 22.85 \, \text{rad/s}$$

The natural period of the structure in this case is:

$$T = \frac{2\pi}{\omega_n} = \frac{2\pi}{22.85} = 0.27 \text{ s}$$

The dynamic response for first phase ($0 \leq t \leq 5$) is:

$$x = \frac{F_o}{k} \times (1 - \cos \omega_n t_d) = \frac{100000}{522057.3} \times (1 - \cos 22.85t)$$
$$= 0.192 \times (1 - \cos 22.85t)$$
$$\dot{x} = \frac{F_o}{k} \times \omega_n \sin \omega_n t_d = \frac{100000}{522057.3} \times 22.85 \times \sin 22.85t$$
$$= 4.38 \times \sin 22.85t$$
$$\ddot{x} = \frac{F_o}{k} \times \omega_n^2 \cos \omega_n t_d = \frac{F_o}{k}\frac{k}{m} \cos \omega_n t_d = \frac{F_o}{m} \cos \omega_n t_d$$
$$= \frac{100000}{1000} \times \cos 22.85t = 100 \times \cos 22.85t$$

The dynamic response for second phase ($5 \leq t \leq 10$) is:

$$x = \frac{F_o}{k}[\cos \omega_n(t - t_d) - \cos \omega_n t] = 0.192 \times [\cos 22.85(t - 5) - \cos 22.85t]$$
$$\dot{x} = \frac{F_o}{k} \times \omega_n[-\sin \omega_n(t - t_d) + \sin \omega_n t]$$
$$= 4.38 \times [-\sin 22.85(t - 5) + \sin 22.85t]$$
$$\ddot{x} = -\frac{F_o}{m}[\cos \omega_n(t - t_d) - \cos \omega_n t] = -100 \times [\cos 22.85(t - 5) - \cos 22.85t]$$

5.3 Earthquake Excitation

The earthquake excitation is considered as an external force which applied to the base of the structure. Since the earthquakes are recorded as ground acceleration in small time interval, the force imposed to the structure due to ground motion can be written based on Newton's second law of motion as:

$$F_e(t) = m\ddot{y}(t)$$

In the equation above, m is the mass of body in kg and $\ddot{y}(t)$ is acceleration due to earthquake, a time-dependent function.

The earthquake acceleration excitation for single degree of freedom system assumes to generate displacement in either vertical or horizontal direction as shown in Fig. 5.19.

The general equation of motion of single degree of freedom system subjected to ground motion is as shown in Eq. (5.8). By subtracting $m\ddot{y}$ from both sides of equation of motion:

$$m\ddot{x} + c(\dot{x} - \dot{y}) + k(x - y) - m\ddot{y} = -m\ddot{y}$$

Rearrange the equation above yields:

$$m(\ddot{x} - \ddot{y}) + c(\dot{x} - \dot{y}) + k(x - y) = -m\ddot{y} \qquad (5.29)$$

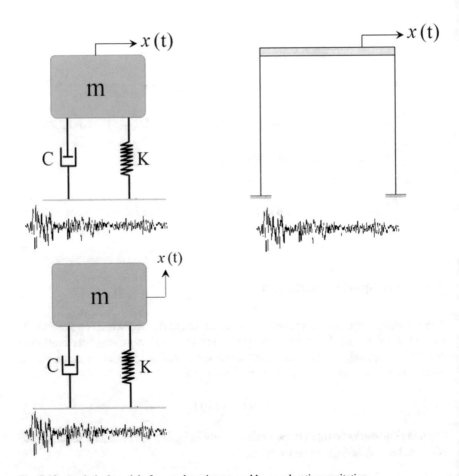

Fig. 5.19 Analytical model of ground motion caused by acceleration excitation

In the equation above, notation x and y are defined as follows:

$$\left.\begin{array}{c} x \\ \dot{x} \\ \ddot{x} \end{array}\right\} \text{Dynamic Response of Mass}$$

$$\left.\begin{array}{c} y \\ \dot{y} \\ \ddot{y} \end{array}\right\} \text{Dynamic Response of Ground}$$

$$\left.\begin{array}{c} x - y \\ \dot{x} - \dot{y} \\ \ddot{x} - \ddot{y} \end{array}\right\} \text{Relative Motion}$$

Let $(x - y) = z$, $(\dot{x} - \dot{y}) = \dot{z}$, $(\ddot{x} - \ddot{y}) = \ddot{z}$, and substitute these into Eq. (5.29) yields:

$$m\ddot{z} + c\dot{z} + kz = -m\ddot{y} \tag{5.30}$$

The equation above describes the dynamic response under force vibration with $F(t) = -m\ddot{y}$. The acceleration due to an earthquake is the external load exerted to structure in this case. For undamped system, the dynamic responses due to earthquake can be determined using Duhamel's integral as shown in Eq. (5.20):

$$z = \frac{1}{m\omega_n} \int_0^t -m\ddot{y}\sin \omega_n(t - \tau)d\tau$$

Simplify the equation above yields:

$$z = \frac{1}{\omega_n} \int_0^t -\ddot{y}\sin \omega_n(t - \tau)d\tau \tag{5.31}$$

For damped system, the expression of load as shown in Eq. (5.20) is transformed as follow to take the damping component into account:

$$x = \frac{1}{m\omega_d} \int_0^t F(\tau)e^{-\xi \omega_n(t-\tau)}\sin \omega_n(t - \tau)d\tau.$$

The dynamic responses due to earthquake is:

$$z_d = \frac{1}{m\omega_d} \int\limits_0^t -m\ddot{y}e^{-\xi\omega_n(t-\tau)}\sin\omega_n(t-\tau)d\tau$$

Simplify the equation above yields:

$$z_d = \frac{1}{\omega_d} \int\limits_0^t -\ddot{y}e^{-\xi\omega_n(t-\tau)}\sin\omega_n(t-\tau)d\tau \tag{5.32}$$

Equations (5.31) and (5.32) give the relative displacement at any instant of time and it depends on the following factors:

(a) Ground acceleration.
(b) Damping of structure system.
(c) Natural period of vibration.

There are two methods to determine the dynamic response in terms of relative displacement, velocity, and acceleration $(z, \dot{z}, \ddot{z})$ of structure system under earthquake excitation which is Duhamel integral and numerical integral.

5.3.1 Spectra Theory

As a focus point of the civil and structural engineers during the design stage of any structure is to check the adequacy of that structure in terms of sheer force, axial force and bending moment.

For the earthquake design domain, the seismic engineer's designers are interested in maximum displacement.

$$S_d : \text{Spectra of displacement}$$
$$S_v : \text{Spectra of velocity}$$
$$S_a : \text{Spectra of acceleration}$$

The displacement spectra, S_d can be expressed using Eq. (5.32):

$$|z_{max}| = S_d = \frac{1}{\omega_d} \int\limits_0^t -\ddot{y}e^{-\xi\omega_n(t-\tau)}\sin\omega_n(t-\tau)d\tau$$

Velocity spectra is related to displacement spectra in the following way:

$$S_d = \frac{1}{\omega_d} S_v$$

When the damping ratio is less than 20%, the value of $\omega_d = \omega_n \sqrt{(1 - \xi^2)} \cong \omega_n$. Therefore,

$$S_d = \frac{1}{\omega_n} S_v \tag{5.33}$$

Consider undamped system subjected to free vibration, Eq. (5.30) is thus transformed as follows:

$$m\ddot{z} + kz = 0$$

Divide both sides of the equation above by m yields:

$$\ddot{z} + \frac{k}{m} z = 0$$

By substituting Eq. (3.14) into equation above yields:

$$\ddot{z} + \omega_n^2 z = 0$$

Express the relative acceleration in terms of other parameters yields:

$$\ddot{z} = \left| -\omega_n^2 z \right|$$

It can be expressed in terms of S_d and S_a as follows:

$$S_a = \omega_n^2 S_d$$

By substituting to Eq. (5.33) into equation above yields:

$$S_a = \omega_n S_v \tag{5.34}$$

The velocity and acceleration spectra as shown in Eqs. (5.33) and (5.34) are often expressed in logarithmic function:

$$\log S_v = \log \omega + \log S_d$$
$$\log S_a = \log \omega + \log S_v$$

Figure 5.20 shows the response spectra for elastic system after ElCentro earthquake as recorded in the year 1940.

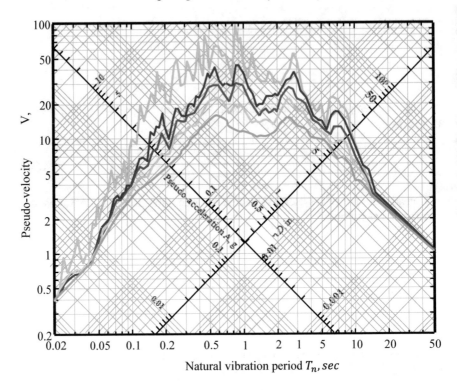

Fig. 5.20 Response spectra for elastic system based on El Centro earthquake recorded in year 1940 (Paz and Leigh 2004)

5.3.2 Base Shear

This approach is an approximate method used in equivalent static design for seismic resistant structure. By Hooke's law,

$$F = kx$$

Base shear force, V_B is correlated to the displacement spectra of structure, S_d. The equation above is thus expressed as:

$$V_B = kS_d$$

By multiplying the terms on the right side of the equation with $\frac{mg}{mg}$ yields:

$$V_B = kS_d \times \frac{mg}{mg}$$

Rearrange the equation above yields:

$$V_B = \frac{k}{m} \times S_d \times \frac{mg}{g}$$

By substituting Eq. (3.14) into the equation above yields:

$$V_B = \omega_n^2 S_d \times \frac{mg}{g}$$

By substituting $S_a = \omega_n^2 S_d$ into equation above yields:

$$V_B = S_a \times \frac{mg}{g}$$

Rearrange the equation above yields:

$$V_B = \frac{S_a}{g} \times mg$$

Let earthquake load coefficient, $C = \frac{S_a}{g}$, and the weight of structure, $W = mg$, the equation above can be rewritten as:

$$V_B = C \times W$$

There are some factors that affect the value of earthquake coefficient, C that is often addressed in various design standard. Some of those factors are:

(a) The importance of structure.
(b) The soil profile.
(c) Non-linearity effect, R, which depends on the structure system type and type of material.

Example 5.5 Earthquake Excitation
The steel frame in Fig. 5.21 was subjected to earthquake excitation, determine the maximum bending stress if the damping ratio is assumed to be 2%.

Solution
By referring to Table A.1 in Appendix, second moment of area, I for rectangular hollow section can be determined using the following equation:

$$I = \frac{bh^3}{12} - \frac{(b - 2t_w)(h - 2t_f)^3}{12}$$
$$= \frac{0.08^4 - (0.08 - 2 \times 0.005)^4}{12} = 1.4125 \times 10^{-6}\,\text{m}^4$$

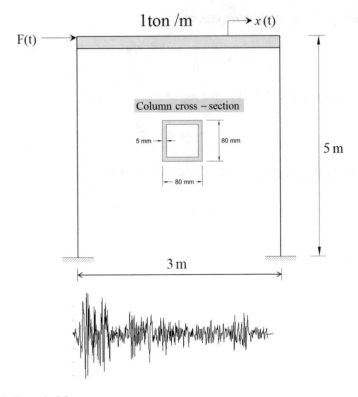

Fig. 5.21 Example 5.5

By referring to Table A.2 in Appendix, stiffness, k for fixed ends column can be determined using the following equation:

$$k = 2 \times \frac{12EI}{L^3}$$
$$= 2 \times \frac{12 \times 200 \times 10^9 \times 1.4125 \times 10^{-4}}{5^3} = 82,382.81 \, \text{N/m}$$

The natural frequency of the structure in this case is:

$$\omega_n = \sqrt{\frac{k}{m}} = \sqrt{\frac{82382.81}{3000}} = 5.24 \, \text{rad/s}$$

The natural period of the structure in this case is:

$$T = \frac{2\pi}{\omega_n} = \frac{2\pi}{5.24} = 1.2 \, \text{s}$$

Based on $T = 1.2\,\text{s}$ and $\xi = 2\%$, by referring to Fig. 5.20, the maximum ground displacement, s_d can be taken as:

$$s_d = 20\,\text{cm} = 200\,\text{mm}$$

Maximum shear force in the beam for this case is:

$$V_{max} = k \times x_{max} = \frac{54240}{2} \times 0.2 = 5424\,\text{N}$$

Maximum bending moment in the beam for this case is:

$$M_{max} = V_{max} \times L = 5424 \times 5 = 27120\,\text{Nm}$$

$$\sigma_{max} = \frac{M \times c}{I} = \frac{27120 \times 0.04}{1.4125 \times 10^{-6}}$$
$$= 768 \times 10^6\,\text{N/m}^2 = 768\,\text{MPa}$$

5.4 Numerical Evaluation of Dynamic Response

There are two methods used to numerically evaluate the dynamic response of structure system: central difference method (CDM) and Newmark's method.

5.4.1 Central Difference Method (CDM)

This method is called explicit method of dynamic analysis which uses the preceding displacement at time t_i to determine the succeeding displacement at time t_{i+1}. Fig. 5.22 shows the concept of central difference method (CDM).

The gradient of the curve, θ is the rate of change of displacement, which is also the velocity.

$$\theta_i = \frac{u_{i+1} - u_i}{\Delta t} \text{ and } \theta_{i-1} = \frac{u_i - u_{i-1}}{\Delta t} \tag{5.35}$$

When the time interval, Δt is constant, the velocity response over the period is:

$$\theta = \dot{u} = \frac{u_{i+1} - u_{i-1}}{2\Delta t} \tag{5.36}$$

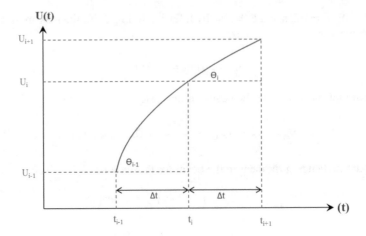

Fig. 5.22 Central difference method (CDM)

Acceleration is the rate of change of velocity which can be expressed as follows:

$$\ddot{u} = \frac{\theta_i - \theta_{i-1}}{\Delta t}$$

By substituting Eq. (5.35) into equation above yields:

$$\ddot{u} = \frac{\frac{u_{i+1} - u_i}{\Delta t} - \frac{u_i - u_{i-1}}{\Delta t}}{\Delta t}$$

$$\ddot{u} = \frac{u_{i+1} - 2u_i + u_{i-1}}{\Delta t^2} \qquad (5.37)$$

Consider the equation of motion:

$$m\ddot{u} + c\dot{u} + ku = f_i$$

By substituting Eqs. (5.36) and (5.37) into equation above yields:

$$m\left[\frac{u_{i+1} - 2u_i + u_{i-1}}{\Delta t^2}\right] + c\left[\frac{u_{i+1} - u_{i-1}}{2\Delta t}\right] + ku_i = f_i$$

Rearrange the equation above yields:

$$\left[\frac{m}{\Delta t^2} + \frac{c}{2\Delta t}\right]u_{i+1} + \left[\frac{m}{\Delta t^2} - \frac{c}{2\Delta t}\right]u_{i-1} + \left[k - \frac{2m}{\Delta t^2}\right]u_i = f_i$$

The purpose of CDM is to predict the displacement response of the structure in the next time step. In this case, force is the function of displacement responses in previous time steps. By rearranging the equation above yields:

$$\left[\frac{m}{\Delta t^2} + \frac{c}{2\Delta t}\right] u_{i+1} = f_i - \left[\frac{m}{\Delta t^2} - \frac{c}{2\Delta t}\right] u_{i-1} - \left[k - \frac{2m}{\Delta t^2}\right] u_i$$

Let $\hat{k} = \left[\frac{m}{\Delta t^2} + \frac{c}{2\Delta t}\right]$, $a = \left[\frac{m}{\Delta t^2} - \frac{c}{2\Delta t}\right]$ and $b = \left[k - \frac{2m}{\Delta t^2}\right]$, the equation above is transformed to:

$$\hat{k} u_{i+1} = f_i - a u_{i-1} - b u_i$$

The equation above can also be written in the form of Hooke's law:

$$\hat{k} u_{i+1} = \hat{F}_i$$

From the equation above

$$\hat{F}_i = f_i - a u_{i-1} - b u_i$$

Under initial condition, $t = 0$, $i = 0$. By substituting the parameters into Eq. (5.36) yields:

$$\dot{u}_o = \frac{u_1 - u_{-1}}{2\Delta t}$$

By rearranging the equation above yields:

$$u_1 = 2\Delta t \dot{u}_o + u_{-1} \tag{5.38}$$

Similarly, Eq. (5.37) can be written as:

$$\ddot{u}_o = \frac{u_1 - 2u_o + u_{-1}}{\Delta t^2} \tag{5.39}$$

By substituting Eq. (5.38) into Eq. (5.39) yields:

$$\ddot{u}_o = \frac{2\Delta t \dot{u}_o + u_{-1} - 2u_o + u_{-1}}{\Delta t^2}$$

Simplify the equation above yields:

$$\ddot{u}_o = \frac{2\Delta t \dot{u}_o - 2u_o + 2u_{-1}}{\Delta t^2}$$

$$\ddot{u}_o \Delta t^2 = 2\Delta t \dot{u}_o - 2u_o + 2u_{-1}$$

Table 5.1 CDM table after step 2

i	t_i (s)	f_i (N)	u_{i-1} (m)	u_i (m)	$\hat{F}_i$ (N)	u_{i+1} (m)
0	t_0		u_{-1}	u_0		
1						
2						

Express u_{-1} in terms of other parameters yields:

$$u_{-1} = u_o - \dot{u}_o \times \Delta t + \ddot{u}_o \times \frac{\Delta t^2}{2}$$

The procedure to conduct the dynamic response evaluation using CDM are as follow:

(1) Determine the time step interval, Δt to be used. To achieve stability condition, $\frac{\Delta t}{T} < \frac{1}{\pi}$ should be obeyed.
(2) Calculate the initial condition for acceleration and displacement (Table 5.1).

$$\ddot{u}_o = \frac{f_i - c\dot{u}_o - ku_o}{m}$$

$$u_{-1} = u_o - \dot{u}_o \times \Delta t + \ddot{u}_o \times \frac{\Delta t^2}{2}$$

(3) Calculate the effective stiffness.

$$\hat{k} = \left[\frac{m}{\Delta t^2} + \frac{c}{2\Delta t} \right]$$

(4) Calculate a and b constants.

$$a = \left[\frac{m}{\Delta t^2} - \frac{c}{2\Delta t} \right], b = \left[k - \frac{2m}{\Delta t^2} \right]$$

(5) Calculate effective load vector (Table 5.2).

$$\hat{F}_i = f_i - au_{i-1} - bu_i$$

Table 5.2 CDM table after step 5

i	t_i (s)	f_i (N)	u_{i-1} (m)	u_i (m)	$\hat{F}_i$ (N)	u_{i+1} (m)
0	t_0	f_0	u_{-1}	u_0	$\hat{F}_0$	
1						
2						

Table 5.3 CDM table after step 6

i	t_i (s)	f_i (N)	u_{i-1} (m)	u_i (m)	$\hat{F}_i$ (N)	u_{i+1} (m)
0	t_0	f_0	u_{-1}	u_0	$\hat{F}_0$	u_1
1						
2						

(6) Calculate the displacement at next time step (Table 5.3).

$$u_{i+1} = \frac{\hat{F}_i}{\hat{k}}$$

(7) Bring over the calculated displacement to the subsequent row as shown in Table 5.4. Repeat step 4 and 5 for each load increment.

Example 5.6 Central Difference Method (CDM)
One-storey four-bay steel frame shown in Fig. 5.23 was subjected to harmonic force vibration with 10 cm initial displacement and zero initial velocity. The horizontal rigid member with mass of 2 ton was supported on four columns having a total stiffness of 400 kN/m and the damping of structure system is found to be 50 kN/m/s. Determine the dynamic response of structure using CDM.

Solution

Table 5.4 CDM table after step 7

i	t_i (s)	f_i (N)	u_{i-1} (m)	u_i (m)	$\hat{F}_i$ (N)	u_{i+1} (m)
0	t_0	f_0	u_{-1}	u_0	$\hat{F}_0$	u_1
1	t_1		u_0	u_1		
2						

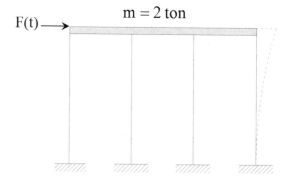

Time (s)	Load (kN)
0	0
0.1	-40
0.2	-100
0.3	-40
0.4	0
0.5	30
0.6	80
0.7	130

Fig. 5.23 Example 5.6

Recommended time step interval is 0.1 s to properly take all applied load into consideration. The natural frequency of structure is:

$$\omega_n = \sqrt{\frac{k}{m}} = \sqrt{\frac{400 \times 10^3}{2000}} = 14.14 \, \text{rad/s}$$

Natural period of the structure is:

$$T = \frac{2\pi}{\omega_n} = \frac{2\pi}{14.14} = 0.444 \, \text{s}$$

The ratio of time step interval to natural period of the structure is:

$$\frac{\Delta t}{T} = \frac{0.1}{0.444} = 0.225 < \frac{1}{\pi} = 0.318$$

Therefore, the recommended time step interval is acceptable.

$$\ddot{u}_o = \frac{f_i - c\dot{u}_o - ku_o}{m} = \frac{0 - 5 \times 10^4 \times 0 - 4 \times 10^5 \times 0.1}{2000} = -20 \, \text{m/s}^2$$

$$u_{-1} = u_o - \dot{u}_o \times \Delta t + \ddot{u}_o \times \frac{\Delta t^2}{2} = 0.1 - 0 \times 0.1 - \frac{20 \times 0.1^2}{2} = 0$$

$$\hat{k} = \left[\frac{m}{\Delta t^2} + \frac{c}{2\Delta t} \right] = \left[\frac{2000}{0.1^2} + \frac{50 \times 10^3}{2 \times 0.1} \right] = 45 \times 10^4 \, \text{N/m}$$

$$a = \left[\frac{m}{\Delta t^2} - \frac{c}{2\Delta t} \right] = \left[\frac{2000}{0.1^2} - \frac{50 \times 10^3}{2 \times 0.1^2} \right] = -5 \times 10^4$$

$$b = \left[k - \frac{2m}{\Delta t^2} \right] = \left[400 \times 10^3 - \frac{2 \times 2000}{0.1^2} \right] = 0$$

At the first time step where $t = t_0 = 0$:

$$\hat{F}_0 = f_0 - au_{-1} - bu_0 = 0 + 5 \times 10^4 \times 0 + 0 \times 0.1 = 0$$

$$u_1 = \frac{\hat{F}_0}{\hat{k}} = \frac{0}{45 \times 10^4} = 0$$

At $t = t_1 = 0.1$ s, the structural response for next instance is again being predicted by using the loading condition at this instance:

$$\hat{F}_1 = f_1 - au_0 - bu_1 = -40 \times 10^3 + 5 \times 10^4 \times 0.1 + 0 \times 0 = -35000$$

Table 5.5 Dynamic response of frame using CDM

i	t_i (s)	f_i (N)	u_{i-1} (m)	u_i (m)	$\hat{F}_i$ (N)	u_{i+1} (m)
0	0	0	0	0.1	0	0
1	0.1	−40000	0.1	0	−35000	−0.08
2	0.2	−100000	0	−0.08	−100000	−0.22
3	0.3	−40000	−0.08	−0.22	−44000	−0.098
4	0.4	0	−0.22	−0.098	−11000	−0.024
5	0.5	30000	−0.098	−0.024	25100	0.06
6	0.6	80000	−0.024	0.06	78800	0.175
7	0.7	130000	0.06	0.175	133000	0.3

$$u_2 = \frac{\hat{F}_1}{\hat{k}} = \frac{-35000}{45 \times 10^4} = -0.08$$

At $t = t_2 = 0.2$ s, the structural response for $t = 0.3$ s is as follows using the value of current displacement predicted during previous iterations:

$$\hat{F}_2 = f_2 - au_1 - bu_2 = -100 \times 10^3 + 5 \times 10^4 \times 0 + 0 \times -0.08 = -100000$$

$$u_3 = \frac{\hat{F}_1}{\hat{k}} = \frac{-100000}{45 \times 10^4} = -0.22$$

At $t = t_3 = 0.3$ s:

$$\hat{F}_3 = f_3 - au_2 - bu_3 = -40 \times 10^3 + 5 \times 10^4 \times -0.08 + 0 \times -0.22 = -44000$$

$$u_4 = \frac{\hat{F}_1}{\hat{k}} = \frac{-44000}{45 \times 10^4} = -0.098$$

The iteration can be continued even after there is no external load being applied to the structure. The purpose of this practice is to inspect the response of structure where effect of damping can be clearly observed. In this example, the calculation is only performed until the last instance of loading where $t = t_7 = 0.7$ s (Table 5.5).

5.5 Exercises

Exercise 5.1 Determine the square hollow section size of steel beam shown in Fig. 5.24. The beam was under the effect of vibration due to generating machine noting that the maximum static displacement of beam is 30 mm downward. Given mass of machine, $m = 25$ ton, mass of moving component, $m' = 25$ kg and it operates

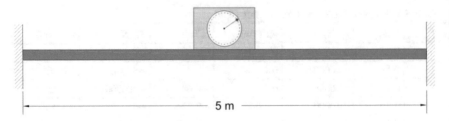

Fig. 5.24 Exercise 5.1

at 250 rpm. Take damping ratio $\xi = 5\%$ and let $e = 250$ mm. The thickness of the beam shall be 10 mm and its mass is to be neglected.

Exercise 5.2 One-storey building, shown in Fig. 5.25 is modelled as 5 m height with two steel columns fixed at the base and a rigid beam supporting 1 ton/m. Determine the maximum response to a triangular impulse of amplitude 5 kN and duration of 1 s. The response of interest is the horizontal displacement at the top of the frame and the bending stress in the columns.

Exercise 5.3 The steel frame in Fig. 5.26 was subjected to earthquake excitation, determine the maximum bending stress if the damping ratio is assumed to be 2%.

Exercise 5.4 Steel frame with rigid horizontal member supporting mass of 5 ton as shown in Fig. 5.27. The frame was subjected to harmonic force vibration. Assume the critical damping ratio of the frame system is 2%. Determine dynamic response of the frame using central difference method. Given that:

(a) Initial displacement $= 0.1$ m
(b) Initial velocity $= 5$ cm/ s
(c) Steel modulus of elasticity 200 GPa

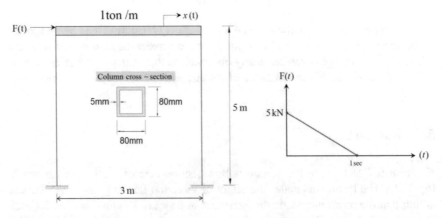

Fig. 5.25 Exercise 5.2

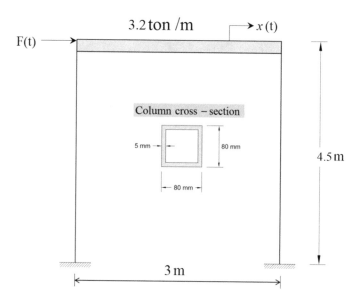

Fig. 5.26 Exercise 5.3

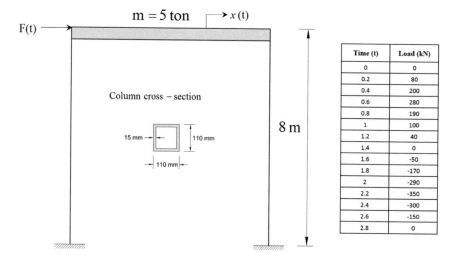

Fig. 5.27 Exercise 5.4

(d) Hollow section 5.110 × 110 mm with thickness = 15 mm.

Chapter 6
Multi-degree of Freedom System

A multi-degree of freedom (MDOF) system describes structural motion with multiple degrees of freedom. This simulates actual structure that consists of multiple storeys. However, the analysis procedure is much complicated than SDOF system because each degree of freedom will have their own displacement, and all of them are related to the response of adjacent degree of freedom (Fig. 6.1).

6.1 Equation of Motion

The formulation of multi-degree of freedom system here is based on equilibrium formulation approach. Figure 6.2 shows the diagram of three-storey building with its analytical model. It is noteworthy that since the adjacent masses are connected to each other, their dynamic response will be related to each other as well.

Based on Fig. 6.2, the motion equation for mass 1 is:

$$m_1\ddot{x}_1 + c_1\dot{x}_1 + k_1x_1 - c_2(\dot{x}_2 - \dot{x}_1) - k_2(x_2 - x_1) = f_1(t)$$

Rearrange the equation above yields:

$$m_1\ddot{x}_1 + (c_1 + c_2)\dot{x}_1 - c_2\dot{x}_2 + (k_1 + k_2)x_1 - k_2x_2 = f_1(t) \qquad (6.1)$$

Based on Fig. 6.2, the motion equation for mass 2 is:

$$m_2\ddot{x}_2 + c_2(\dot{x}_2 - \dot{x}_1) + k_2(x_2 - x_1) - c_3(\dot{x}_3 - \dot{x}_2) - k_3(x_3 - x_2) = f_2(t)$$

© The Editor(s) (if applicable) and The Author(s), under exclusive license to Springer Nature Singapore Pte Ltd. 2020
F. Hejazi and T. K. Chun, *Conceptual Theories in Structural Dynamics*,
Advanced Structured Materials 135,
https://doi.org/10.1007/978-981-15-5440-7_6

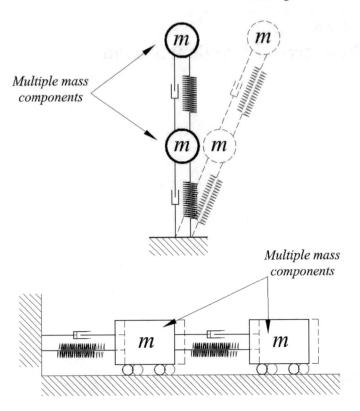

Fig. 6.1 Analytical models for MDOF system

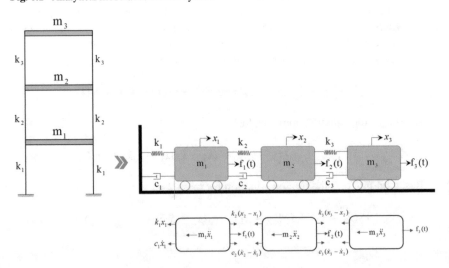

Fig. 6.2 Three-storey building and its MDOF analytical model

Rearrange the equation above yields:

$$m_2\ddot{x}_2 - c_2\dot{x}_1 + (c_2 + c_3)\dot{x}_2 - c_3\dot{x}_3 - k_2x_1 + (k_2 + k_3)x_2 - k_3x_3 = f_2(t) \quad (6.2)$$

Based on Fig. 6.2, the motion equation for mass 3 is:

$$m_3\ddot{x}_3 + c_3(\dot{x}_3 - \dot{x}_2) + k_3(x_3 - x_2) = f_2(t)$$

Rearrange the equation above yields:

$$m_2\ddot{x}_3 + c_3\dot{x}_2 - c_3\dot{x}_3 - k_3x_2 + k_3x_3 = f_2(t) \quad (6.3)$$

Express Eqs. (6.1), (6.2) and (6.3) in matrix form yields:

$$
\begin{bmatrix} m_1 & 0 & 0 \\ 0 & m_2 & 0 \\ 0 & 0 & m_3 \end{bmatrix}
\begin{bmatrix} \ddot{x}_1 \\ \ddot{x}_2 \\ \ddot{x}_3 \end{bmatrix}
+
\begin{bmatrix} c_1 + c_2 & -c_2 & 0 \\ -c_2 & c_2 + c_3 & -c_3 \\ 0 & -c_3 & c_3 \end{bmatrix}
\begin{bmatrix} \dot{x}_1 \\ \dot{x}_2 \\ \dot{x}_3 \end{bmatrix}
$$

$$
+
\begin{bmatrix} k_1 + k_2 & -k_2 & 0 \\ -k_2 & k_2 + k_3 & -k_3 \\ 0 & -k_3 & k_3 \end{bmatrix}
\begin{bmatrix} x_1 \\ x_2 \\ x_3 \end{bmatrix}
$$

$$
=
\begin{bmatrix} f_1(t) \\ f_2(t) \\ f_3(t) \end{bmatrix}
$$

Generally, for n-degree of freedom system, the above matrix equation can be generalized as follows:

$$[m]_{n \times n}\{\ddot{x}(t)\}_{n \times 1} + [c]_{n \times n}\{\dot{x}(t)\}_{n \times 1} + [k]_{n \times n}\{x(t)\}_{n \times 1} = \{f(t)\}_{n \times 1}$$

$[m]$, $[c]$ and $[k]$ are symmetrical matrices where the axis of symmetry is non-zero. The general form of these matrices are as follows:

$$
[m] =
\begin{bmatrix}
m_1 & 0 & 0 & 0 & 0 & 0 \\
0 & m_2 & 0 & 0 & 0 & 0 \\
0 & 0 & m_3 & 0 & 0 & 0 \\
0 & 0 & 0 & \vdots & \vdots & \vdots \\
0 & 0 & 0 & \dots & m_{n-1} & 0 \\
0 & 0 & 0 & \dots & 0 & m_n
\end{bmatrix}
$$

$$[c] = \begin{bmatrix} c_1 + c_2 & -c_2 & 0 & 0 & 0 & 0 \\ -c_2 & c_2 + c_3 & -c_3 & 0 & 0 & 0 \\ 0 & -c_3 & c_3 + c_4 & -c_4 & 0 & 0 \\ 0 & 0 & -c_4 & \vdots & \vdots & \vdots \\ 0 & 0 & 0 & \cdots & c_{n-1} + c_n & -c_n \\ 0 & 0 & 0 & \cdots & -c_n & c_n \end{bmatrix}$$

$$[k] = \begin{bmatrix} k_1 + k_2 & -k_2 & 0 & 0 & 0 & 0 \\ -k_2 & k_2 + k_3 & -k_3 & 0 & 0 & 0 \\ 0 & -k_3 & k_3 + k_4 & -k_4 & 0 & 0 \\ 0 & 0 & -k_4 & \vdots & \vdots & \vdots \\ 0 & 0 & 0 & \cdots & k_{n-1} + k_n & -k_n \\ 0 & 0 & 0 & \cdots & -k_n & k_n \end{bmatrix} \tag{6.4}$$

6.2 Multi-degree of Freedom for Undamped System Under Free Vibration

By considering the undamped system under free vibration for two-degree of freedom (2-DOF) system as shown in Fig. 6.3.

Based on Fig. 6.3, the equation of motion for mass 1 is:

$$m_1\ddot{x}_1 + k_1 x_1 - k_2(x_2 - x_1) = 0$$

Rearrange the equation above yields:

$$m_1\ddot{x}_1 + (k_1 + k_2)x_1 - k_2 x_2 = 0 \tag{6.5}$$

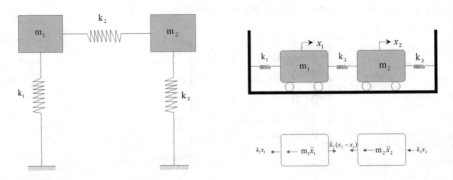

Fig. 6.3 Undamped 2-DOF system under free vibration

Based on Fig. 6.3, the equation of motion for mass 2 is:

$$m_2\ddot{x}_2 + k_2(x_2 - x_1) + k_3x_2 = 0$$

Rearrange the equation above yields:

$$m_2\ddot{x}_2 - k_2x_1 + (k_2 + k_3)x_2 = 0 \tag{6.6}$$

Express Eqs. (6.5) and (6.6) in matrix form yields:

$$\begin{bmatrix} m_1 & 0 \\ 0 & m_2 \end{bmatrix}\begin{bmatrix} \ddot{x}_1 \\ \ddot{x}_2 \end{bmatrix} + \begin{bmatrix} (k_1 + k_2) & -k_2 \\ -k_2 & (k_2 + k_3) \end{bmatrix}\begin{bmatrix} x_1 \\ x_2 \end{bmatrix} = \begin{bmatrix} 0 \\ 0 \end{bmatrix} \tag{6.7}$$

Let the following be the trial answer for the dynamic response:

$$x_1 = a_1 \sin \omega_n t, \ \dot{x}_1 = a_1\omega_n \cos \omega_n t, \ \ddot{x}_1 = -a_1\omega_n^2 \sin \omega_n t$$
$$x_2 = a_2 \sin \omega_n t, \ \dot{x}_2 = a_2\omega_n \cos \omega_n t, \ \ddot{x}_2 = -a_2\omega_n^2 \sin \omega_n t \tag{6.8}$$

By substituting the trial answers in Eq. (6.8) into Eq. (6.7) yields:

$$\begin{bmatrix} m_1 & 0 \\ 0 & m_2 \end{bmatrix}\begin{bmatrix} -a_1\omega_n^2 \sin \omega_n t \\ -a_2\omega_n^2 \sin \omega_n t \end{bmatrix} + \begin{bmatrix} (k_1 + k_2) & -k_2 \\ -k_2 & (k_2 + k_3) \end{bmatrix}\begin{bmatrix} a_1 \sin \omega_n t \\ a_2 \sin \omega_n t \end{bmatrix} = \begin{bmatrix} 0 \\ 0 \end{bmatrix}$$

By eliminating the common term $\sin \omega_n t$ from matrix above yields:

$$\begin{bmatrix} m_1 & 0 \\ 0 & m_2 \end{bmatrix}\begin{bmatrix} -a_1\omega_n^2 \\ -a_2\omega_n^2 \end{bmatrix} + \begin{bmatrix} (k_1 + k_2) & -k_2 \\ -k_2 & (k_2 + k_3) \end{bmatrix}\begin{bmatrix} a_1 \\ a_2 \end{bmatrix} = \begin{bmatrix} 0 \\ 0 \end{bmatrix}$$

Expand the above yields:

$$\begin{bmatrix} -m_1\omega_n^2 & 0 \\ 0 & -m_2\omega_n^2 \end{bmatrix}\begin{bmatrix} -a_1 \\ -a_2 \end{bmatrix} + \begin{bmatrix} (k_1 + k_2) & -k_2 \\ -k_2 & (k_2 + k_3) \end{bmatrix}\begin{bmatrix} a_1 \\ a_2 \end{bmatrix} = \begin{bmatrix} 0 \\ 0 \end{bmatrix}$$

Simplify the matrix above yields:

$$\left(\begin{bmatrix} -m_1\omega_n^2 & 0 \\ 0 & -m_2\omega_n^2 \end{bmatrix} + \begin{bmatrix} (k_1 + k_2) & -k_2 \\ -k_2 & (k_2 + k_3) \end{bmatrix}\right)\begin{bmatrix} a_1 \\ a_2 \end{bmatrix} = \begin{bmatrix} 0 \\ 0 \end{bmatrix}$$
$$\begin{bmatrix} k_1 + k_2 - m_1\omega_n^2 & -k_2 \\ -k_2 & k_2 + k_3 - m_2\omega_n^2 \end{bmatrix}\begin{bmatrix} a_1 \\ a_2 \end{bmatrix} = \begin{bmatrix} 0 \\ 0 \end{bmatrix} \tag{6.9}$$

Since a_1 and a_2 are the solution for the equation, they cannot be zero. Therefore,

$$\det \begin{bmatrix} k_1 + k_2 - m_1\omega_n^2 & -k_2 \\ -k_2 & k_2 + k_3 - m_2\omega_n^2 \end{bmatrix} = 0$$

Based on the expression above, the following can be derived:

$$\left[(k_1 + k_2) - m_1\omega_n^2\right] \times \left[(k_2 + k_3) - m_2\omega_n^2\right] - k_2^2 = 0$$

Expand the equation above yields:

$$(k_1 + k_2)(k_2 + k_3) - m_2\omega_n^2(k_1 + k_2) - m_1\omega_n^2(k_2 + k_3) + m_1m_2\omega_n^4 - k_2^2 = 0$$
$$k_1k_2 + k_1k_3 + k_2^2 + k_2k_3 - [m_1(k_2 + k_3) + m_2(k_1 + k_2)]\omega_n^2 + m_1m_2\omega_n^4 - k_2^2 = 0$$

Rearrange the equation above yields:

$$m_1m_2\omega_n^4 - [m_1(k_2 + k_3) + m_2(k_1 + k_2)]\omega_n^2 + k_1k_2 + k_1k_3 + k_2k_3 = 0$$

By dividing the terms in equation above with m_1m_2 yields:

$$\omega_n^4 - \left[\frac{(k_2 + k_3)}{m_2} + \frac{(k_1 + k_2)}{m_1}\right]\omega_n^2 + \frac{k_1k_2 + k_1k_3 + k_2k_3}{m_1m_2} = 0$$

By assuming $m_1 = m_2 = m$ and $k_1 = k_2 = k_3 = k$, the equation above can be simplified into:

$$\omega_n^4 - \left[\frac{2k}{m} + \frac{2k}{m}\right]\omega_n^2 + \frac{3k^2}{m^2} = 0$$
$$\omega_n^4 - \frac{4k}{m}\omega_n^2 + \frac{3k^2}{m^2} = 0$$

Let $\lambda = \omega_n^2$ and $\lambda^2 = \omega_n^4$, the equation above can be rewritten as:

$$\lambda^2 - \frac{4k}{m}\lambda + \frac{3k^2}{m^2} = 0$$

Solve the equation by determining the solution for quadratic equation using $\lambda_{1,2} = \frac{-b \pm \sqrt{b^2 - 4ac}}{2a}$ yields:

$$a = 1, b = -\frac{4k}{m}, c = \frac{3k^2}{m^2}$$

Therefore, the solution to the quadratic equation is:

$$\lambda_{1,2} = \frac{\frac{4k}{m} \pm \sqrt{\left(\frac{4k}{m}\right)^2 - \frac{12k^2}{m^2}}}{2}$$

$$\lambda_{1,2} = \frac{\frac{4k}{m} \pm \sqrt{\frac{16k^2}{m^2} - \frac{12k^2}{m^2}}}{2}$$

$$\lambda_{1,2} = \frac{2k}{m} \pm \frac{k}{m}$$

From the expression above, the following are determined:

$$\lambda_1 = \omega_{n1}^2 = \frac{k}{m}, \; \omega_{n1} = \sqrt{\frac{k}{m}}$$

$$\lambda_2 = \omega_{n2}^2 = \frac{3k}{m}, \; \omega_{n2} = \sqrt{\frac{3k}{m}} \qquad (6.10)$$

The existence of two different values of ω_n denotes a 2-DOF system can exhibit two different forms of vibration. Each form of vibration is known as normal mode. For every mode, corresponding ω_n represents its natural frequency. The first mode of vibration is always the most critical one as all masses displace in the same direction and result in maximum displacement at the top of building. Therefore, the natural frequency for first mode of vibration is also known as fundamental frequency. The fundamental period of vibration can be determined using:

$$T_1 = \frac{2\pi}{\omega_{n1}}$$

Based on the assumption $m_1 = m_2 = m$ and $k_1 = k_2 = k_3 = k$, Eq. (6.9) is simplified:

$$\begin{bmatrix} 2k - m\omega_n^2 & -k \\ -k & 2k - m\omega_n^2 \end{bmatrix} \begin{bmatrix} a_1 \\ a_2 \end{bmatrix} = \begin{bmatrix} 0 \\ 0 \end{bmatrix} \qquad (6.11)$$

By substituting the $\omega_{n1}^2 = \frac{k}{m}$ for first mode of vibration from Eq. (6.10) into equation above yields:

$$\begin{bmatrix} 2k - m\frac{k}{m} & -k \\ -k & 2k - m\frac{k}{m} \end{bmatrix} \begin{bmatrix} a_1 \\ a_2 \end{bmatrix} = \begin{bmatrix} 0 \\ 0 \end{bmatrix}$$

Simplify the above yields:

$$\begin{bmatrix} k & -k \\ -k & k \end{bmatrix} \begin{bmatrix} a_1 \\ a_2 \end{bmatrix} = \begin{bmatrix} 0 \\ 0 \end{bmatrix}$$

From the matrix, the following can be written:

$$ka_1 - ka_2 = 0$$

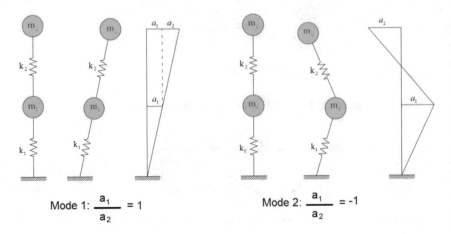

Mode 1: $\dfrac{a_1}{a_2} = 1$ Mode 2: $\dfrac{a_1}{a_2} = -1$

Fig. 6.4 Normal modes of vibration

$$a_1 = a_2$$

By substituting the $\omega_{n2}^2 = \frac{3k}{m}$ for second mode of vibration from Eq. (6.10) into equation above yields:

$$\begin{bmatrix} 2k - m\frac{3k}{m} & -k \\ -k & 2k - m\frac{3k}{m} \end{bmatrix} \begin{bmatrix} a_1 \\ a_2 \end{bmatrix} = \begin{bmatrix} 0 \\ 0 \end{bmatrix}$$

Simplify the above yields:

$$\begin{bmatrix} -k & -k \\ -k & -k \end{bmatrix} \begin{bmatrix} a_1 \\ a_2 \end{bmatrix} = \begin{bmatrix} 0 \\ 0 \end{bmatrix}$$

From the matrix, the following equation can be produced:

$$- ka_1 - ka_2 = 0$$
$$a_1 = -a_2$$

Based on the outcome above, the normal modes for the 2-DOF system under free vibration can be visualized as follows (Fig. 6.4).

6.3 Normalization of Vibration

The dynamic response of structure can be obtained through experiment. However, given the situation where the experiment is carried out with a specific setup, such

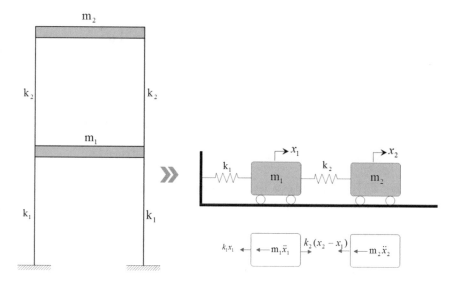

Fig. 6.5 Two-storey building as undamped 2-DOF system under free vibration

result needs to be adjusted in order to be implemented for further analysis. The experimental vibration modes will require adjustment in terms of scaling, a procedure which is commonly known as normalization of vibration. With normalization, the normal modes can be estimated for other similar structures with same mass and stiffness proportion for all degree of freedom.

Consider two-storey undamped frame system subjected to free vibration as shown in Fig. 6.5.

Based on Fig. 6.5, the equation of motion for mass 1 is:

$$m_1\ddot{x}_1 + k_1 x_1 - k_2(x_2 - x_1) = 0$$

Rearrange the equation above yields:

$$m_1\ddot{x}_1 + (k_1 + k_2)x_1 - k_2 x_2 = 0 \qquad (6.12)$$

Based on Fig. 6.5, the equation of motion for mass 2 is:

$$m_2\ddot{x}_2 + k_2(x_2 - x_1) = 0$$

Rearrange the equation above yields:

$$m_2\ddot{x}_2 - k_2 x_1 + k_2 x_2 = 0 \qquad (6.13)$$

Express Eqs. (6.12) and (6.13) in matrix form yields:

$$\begin{bmatrix} m_1 & 0 \\ 0 & m_2 \end{bmatrix} \begin{bmatrix} \ddot{x}_1 \\ \ddot{x}_2 \end{bmatrix} + \begin{bmatrix} (k_1 + k_2) & -k_2 \\ -k_2 & k_2 \end{bmatrix} \begin{bmatrix} x_1 \\ x_2 \end{bmatrix} = \begin{bmatrix} 0 \\ 0 \end{bmatrix}$$

Let the following be the trial answer for the dynamic response:

$$x_1 = a_1 \sin \omega_n t, \ \dot{x}_1 = a_1 \omega_n \cos \omega_n t, \ \ddot{x}_1 = -a_1 \omega_n^2 \sin \omega_n t$$

$$x_2 = a_2 \sin \omega_n t, \ \dot{x}_2 = a_1 \omega_n \cos \omega_n t, \ \ddot{x}_2 = -a_2 \omega_n^2 \sin \omega_n t \qquad (6.14)$$

By substituting the trial answers in Eq. (6.8) into Eq. (6.7) yields:

$$\begin{bmatrix} m_1 & 0 \\ 0 & m_2 \end{bmatrix} \begin{bmatrix} -a_1 \omega_n^2 \sin \omega_n t \\ -a_2 \omega_n^2 \sin \omega_n t \end{bmatrix} + \begin{bmatrix} (k_1 + k_2) & -k_2 \\ -k_2 & k_2 \end{bmatrix} \begin{bmatrix} a_1 \sin \omega_n t \\ a_2 \sin \omega_n t \end{bmatrix} = \begin{bmatrix} 0 \\ 0 \end{bmatrix}$$

By eliminating the common term $\sin \omega_n t$ from matrix above yields:

$$\begin{bmatrix} m_1 & 0 \\ 0 & m_2 \end{bmatrix} \begin{bmatrix} -a_1 \omega_n^2 \\ -a_2 \omega_n^2 \end{bmatrix} + \begin{bmatrix} (k_1 + k_2) & -k_2 \\ -k_2 & k_2 \end{bmatrix} \begin{bmatrix} a_1 \\ a_2 \end{bmatrix} = \begin{bmatrix} 0 \\ 0 \end{bmatrix}$$

Expand the above yields:

$$\begin{bmatrix} -m_1 \omega_n^2 & 0 \\ 0 & -m_2 \omega_n^2 \end{bmatrix} \begin{bmatrix} -a_1 \\ -a_2 \end{bmatrix} + \begin{bmatrix} (k_1 + k_2) & -k_2 \\ -k_2 & k_2 \end{bmatrix} \begin{bmatrix} a_1 \\ a_2 \end{bmatrix} = \begin{bmatrix} 0 \\ 0 \end{bmatrix}$$

Simplify the matrix above yields:

$$\left(\begin{bmatrix} -m_1 \omega_n^2 & 0 \\ 0 & -m_2 \omega_n^2 \end{bmatrix} + \begin{bmatrix} (k_1 + k_2) & -k_2 \\ -k_2 & k_2 \end{bmatrix} \right) \begin{bmatrix} a_1 \\ a_2 \end{bmatrix} = \begin{bmatrix} 0 \\ 0 \end{bmatrix}$$

$$\begin{bmatrix} k_1 + k_2 - m_1 \omega_n^2 & -k_2 \\ -k_2 & k_2 - m_2 \omega_n^2 \end{bmatrix} \begin{bmatrix} a_1 \\ a_2 \end{bmatrix} = \begin{bmatrix} 0 \\ 0 \end{bmatrix} \qquad (6.15)$$

The normalization of mode of vibration can be performed by applying the following formula. This method is also known as mass-normalization, a method that operates by obeying the orthogonality of eigenvectors.

$$\phi_{ij} = \frac{a_{ij}}{\sqrt{\sum_{k=1}^{n} m_k a_{ij}^2}} \qquad (6.16)$$

ϕ_{ij} is the normalized i component of the j modal vector. For example, ϕ_{11} denotes the normalized component for mass 1 under first mode of vibration, while ϕ_{12} denotes the normalized component for mass 1 under second mode of vibration.

$$
\begin{array}{cc}
mode1 & mode2 \\
a_{11} & a_{12} \quad mass1 \\
a_{21} & a_{22} \quad mass2
\end{array}
$$

$$
\begin{array}{cc}
mode1 & mode2 \\
\phi_{11} & \phi_{12} \quad mass1 \\
\phi_{21} & \phi_{22} \quad mass2
\end{array}
$$

Say, $x(t)$, $\dot{x}(t)$ and $\ddot{x}(t)$ are structural response obtained from experiment. To ease the analysis of similar structure in future, these responses are usually being normalized using scale factor ϕ. The modal responses, say $z(t)$, $\dot{z}(t)$ and $\ddot{z}(t)$ define the normal modes of structure, but it will need a revert action to predict the actual magnitude of response. In this case, those modal responses will need to be multiplied with the factor ϕ.

$$
x = \phi z \tag{6.17}
$$

Such normalization is also possible for external force, which will transform it into modal force as a result. Using similar fashion, the external force F relates to modal force P in the following way:

$$
F = \phi P \tag{6.18}
$$

Work done is defined as the product of force and resultant displacement. In matrix form, $\{x\} = \begin{bmatrix} x_1 \\ x_2 \\ x_3 \end{bmatrix}$ and $\{F\} = \begin{bmatrix} F_1 \\ F_2 \\ F_3 \end{bmatrix}$. By using virtual work method, the following can be defined:

$$
W = \{x\}^{\mathrm{T}}\{F\} = \begin{bmatrix} x_1 & x_2 & x_3 \end{bmatrix} \begin{bmatrix} F_1 \\ F_2 \\ F_3 \end{bmatrix}
$$

Also, work done is the same for modal force and displacement regardless of the magnitude of scale being applied. Therefore:

$$
W = \{z\}^{\mathrm{T}}\{P\} = \begin{bmatrix} z_1 & z_2 & z_3 \end{bmatrix} \begin{bmatrix} P_1 \\ P_2 \\ P_3 \end{bmatrix} = \{x\}^{\mathrm{T}}\{F\}
$$

By substituting Eq. (6.17) into the equation above yields:

$$
z^{\mathrm{T}} P = (\phi z)^{\mathrm{T}} F
$$

By expanding the equation above yields:

$$z^T P = z^T \phi^T F$$

By removing the common term z^T from both sides of equation above:

$$P = \phi^T F \qquad (6.19)$$

6.4   Orthogonality of Eigenvectors

Consider undamped equation of motion for multi-degree of freedom:

$$m\ddot{x} + kx = 0 \qquad (6.20)$$

Let the answer to the structural response be:

$$x = \phi_i \sin \omega_n t, \; \dot{x} = \omega_n \phi_i \cos \omega_n t, \; \ddot{x} = -\omega_n^2 \phi_i \sin \omega_n t \qquad (6.21)$$

By substituting the relationships in Eq. (6.21) into Eq. (6.20) yields:

$$-\omega_n^2 \phi_i \sin \omega_n t m + \phi_i \sin \omega_n t k = 0$$

By eliminating the common term $\sin \omega_n t$ from the equation above yields:

$$-\omega_n^2 \phi_i m + \phi_i k = 0$$

In general form for multi-degree of freedom system, the parameters m, k and ϕ_i will need to be expressed in matrix form to take all the degree of freedom into consideration:

$$- \omega_n^2 [m]\{\phi_i\} + [k]\{\phi_i\} = 0$$
$$\omega_n^2 [m]\{\phi_i\} = [k]\{\phi_i\} \qquad (6.22)$$

Equation above should be set for eigenvector s, and by multiplying both sides of the equation by $[\phi_r]^T$ yields:

$$\omega_s^2 [\phi_r]^T [m][\phi_s] = [\phi_r]^T [k][\phi_s]$$

The mass and stiffness are symmetrical properties, which means $[k]^T = [k]$ and $[m]^T = [m]$. Therefore, by transposing the equation above yields:

$$\omega_s^2 [\phi_r][m][\phi_s]^T = [\phi_r][k][\phi_s]^T \qquad (6.23)$$

Also, Eq. (6.22) should be set for eigenvector r, and by multiplying both sides of the equation by $[\phi_s]^T$ yields:

$$\omega_r^2 [\phi_s]^T [m][\phi_r] = [\phi_s]^T [k][\phi_r]$$

Rearrange the equation above yields:

$$\omega_r^2 [\phi_r][m][\phi_s]^T = [\phi_r][k][\phi_s]^T \tag{6.24}$$

By subtracting Eq. (6.24) from Eq. (6.23) yields:

$$\omega_s^2 [\phi_r][m][\phi_s]^T - \omega_r^2 [\phi_r][m][\phi_s]^T = 0$$

By simplifying the equation above yields:

$$\left(\omega_s^2 - \omega_r^2\right)[\phi_r][m][\phi_s]^T = 0$$

Since $\left(\omega_s^2 - \omega_r^2\right) \neq 0$,

$$[\phi_r][m][\phi_s]^T = 0$$

For example, the orthogonality correlation for a two-degree of freedom system can be written as,

$$\sum m_i \times \phi_{ir} \times \phi_{is} = 0$$

In the equation above, i is the number of inspected vibration mode, while r and s are the notation representing all possible number of modes for a system (which is 1 and 2 respectively for two-degree of freedom system).

Generally, for every degree of freedom there will be one mode of vibration. Therefore, for n-degree of freedom system, the orthogonality correlation can be generalized as:

$$\sum m_i \times \phi_{i1} \times \phi_{i2} \times \phi_{i3} \times \cdots \times \phi_{i(n-1)} \times \phi_{in} = 0$$

Example 6.1 Natural Frequency and Normal Modes
Find the natural frequency and normal modes for a reinforced concrete building consists of three typical storeys as shown in Fig. 6.6. All columns are having same dimension, although the reinforcement varies with storey and location. Consider the foundation as fixed support.

Solution

Considering the content of rebar inside column cross-section is insignificant in influencing the lateral stiffness of the entire structure, the stiffness provided by the steel

Fig. 6.6 Example 6.1

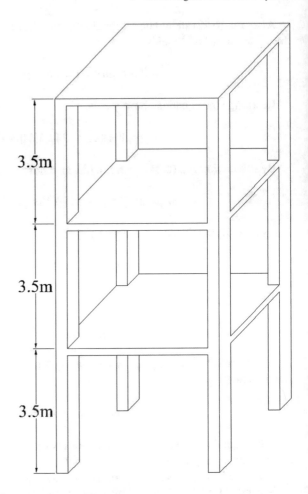

3.5m

3.5m

3.5m

reinforcement can be neglected. Therefore, the stiffness component will only be calculated based on the geometry of concrete column itself.

The position of the columns is the same for all storeys, and the size of columns is the same regardless of its position. Therefore the stiffness component for each storey is the same and thus:

$$k = k_1 = k_2 = k_3$$

The floor elements, i.e. structural and non-structural elements are the same for typical storeys. Thus:

$$m = m_1 = m_2 = m_3$$

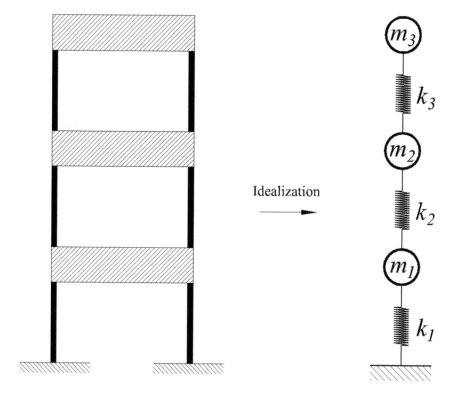

Fig. 6.7 Idealization of structure for Example 6.1

The building can be idealized as three-degree of freedom system as shown in figure below (Fig. 6.7):

The mass and stiffness matrices of the structure are as follow:

$$[m] = \begin{bmatrix} m_1 & 0 & 0 \\ 0 & m_2 & 0 \\ 0 & 0 & m_3 \end{bmatrix} = \begin{bmatrix} m & 0 & 0 \\ 0 & m & 0 \\ 0 & 0 & m \end{bmatrix}$$

$$[k] = \begin{bmatrix} k_1 + k_2 & -k_2 & 0 \\ -k_2 & k_2 + k_3 & -k_3 \\ 0 & -k_3 & k_3 \end{bmatrix} = \begin{bmatrix} 2k & -k & 0 \\ -k & 2k & -k \\ 0 & -k & k \end{bmatrix}$$

By substituting the above into equation of motion for free vibration of undamped system yields:

$$[m]\{\ddot{x}\} + [k]\{x\} = 0 \tag{6.25}$$

Under harmonic vibration, the answer $\{x\}$ can be expressed in sine function and $\{\ddot{x}\}$ can be obtained by taking second-order derivative of $\{x\}$ with respect to time:

$$\{x\} = A \sin \omega_n t$$

$$\{\ddot{x}\} = -A\omega_n^2 \sin \omega_n t$$

In the solution above, A is the amplitude of displacement and $A = \begin{Bmatrix} a_1 \\ a_2 \\ a_3 \end{Bmatrix}$, ω_n is the natural frequency of structure that needs to be solved in this example.

Therefore, by substituting the answer into Eq. (6.25):

$$[m] \times -A\omega_n^2 \sin \omega_n t + [k] \times A \sin \omega_n t = 0$$

By removing the common terms $\sin \omega_n t$ and rearranging it, the following equation obtained:

$$(-\omega_n^2[m] + [k])A = 0 \tag{6.26}$$

Since A is the solution and it is non-zero,

$$-\omega_n^2[m] + [k] = 0$$

By substituting the mass and stiffness matrix into equation above yields:

$$-\omega_n^2 \begin{bmatrix} m & 0 & 0 \\ 0 & m & 0 \\ 0 & 0 & m \end{bmatrix} + \begin{bmatrix} 2k & -k & 0 \\ -k & 2k & -k \\ 0 & -k & k \end{bmatrix} = 0$$

Perform arithmetic addition to the matrices yields:

$$\begin{bmatrix} -m\omega_n^2 + 2k & -k & 0 \\ -k & -m\omega_n^2 + 2k & -k \\ 0 & -k & -m\omega_n^2 + k \end{bmatrix} = 0 \tag{6.27}$$

Determinant of the matrix above must be zero to make the expression valid. Therefore,

$$(-m\omega_n^2 + 2k)\left[(-m\omega_n^2 + 2k)(-m\omega_n^2 + k) - (-k)(-k)\right]$$
$$- (-k)\left[(-k)(-m\omega_n^2 + k)\right] = 0$$

By expanding the equation above:

$$(-m\omega_n^2 + 2k)\left[m^2\omega_n^4 - 3km\omega_n^2 + k^2\right] + k\left[(km\omega_n^2 - k^2)\right] = 0$$
$$- m^3\omega_n^6 + 3km^2\omega_n^4 - k^2m\omega_n^2 + 2km^2\omega_n^4 - 6k^2m\omega_n^2 + 2k^3 + k^2m\omega_n^2 - k^3 = 0$$

By simplifying and rearranging the equation above:

$$-m^3\omega_n^6 + 5km^2\omega_n^4 - 6k^2m\omega_n^2 + k^3 = 0$$

Let $\lambda = \omega_n^2$,

$$-m^3\lambda^3 + 5km^2\lambda^2 - 6k^2m\lambda + k^3 = 0 \tag{6.28}$$

Newton–Raphson method is used to estimate one of the roots for the cubic equation. This is an iterative method that determines the root using trial and error method until convergence is achieved:

$$x_{n+1} = x_n - \frac{f(x_n)}{f'(x_n)}$$

The derivative for Eq. (6.28) is as follows:

$$f'(\lambda) = -3m^3\lambda^2 + 10km^2\lambda - 6k^2m$$

For this example, the general formula for Newton–Raphson method is as follows, where λ_T denotes trial root:

$$\lambda_{T,n+1} = \lambda_{T,n} - \frac{-m^3\lambda^3 + 5km^2\lambda^2 - 6k^2m\lambda + k^3}{-3m^3\lambda^2 + 10km^2\lambda - 6k^2m}$$

Use $\lambda_T = \frac{k}{m}$ for the first trial,

$$\lambda_{T,1} = \frac{k}{m} - \frac{-m^3\left(\frac{k}{m}\right)^3 + 5km^2\left(\frac{k}{m}\right)^2 - 6k^2m\left(\frac{k}{m}\right) + k^3}{-3m^3\left(\frac{k}{m}\right)^2 + 10km^2\left(\frac{k}{m}\right) - 6k^2m}$$

$$= \frac{k}{m} - \frac{-k^3 + 5k^3 - 6k^3 + k^3}{-3k^2m + 10k^2m - 6k^2m} = \frac{k}{m} - \frac{-k^3}{k^2m} = 2\frac{k}{m}$$

For the second trial, $\lambda_T = 2\frac{k}{m}$ is used

$$\lambda_{T,2} = 2\frac{k}{m} - \frac{-m^3\left(2\frac{k}{m}\right)^3 + 5km^2\left(2\frac{k}{m}\right)^2 - 6k^2m\left(2\frac{k}{m}\right) + k^3}{-3m^3\left(2\frac{k}{m}\right)^2 + 10km^2\left(2\frac{k}{m}\right) - 6k^2m}$$

$$= 2\frac{k}{m} - \frac{-8k^3 + 20k^3 - 12k^3 + k^3}{-12k^2m + 20k^2m - 6k^2m} = 2\frac{k}{m} - \frac{k^3}{2k^2m} = 1.5\frac{k}{m}$$

For the third trial, $\lambda_T = 1.5\frac{k}{m}$ is used

$$\lambda_{T,3} = 1.5\frac{k}{m} - \frac{-m^3\left(1.5\frac{k}{m}\right)^3 + 5km^2\left(1.5\frac{k}{m}\right)^2 - 6k^2m\left(1.5\frac{k}{m}\right) + k^3}{-3m^3\left(1.5\frac{k}{m}\right)^2 + 10km^2\left(1.5\frac{k}{m}\right) - 6k^2m}$$

$$= 1.5\frac{k}{m} - \frac{-3.375k^3 + 11.25k^3 - 9k^3 + k^3}{-6.75k^2m + 15k^2m - 6k^2m} = 1.5\frac{k}{m} - \frac{-0.125k^3}{2.25k^2m}$$

$$= 1.56\frac{k}{m}$$

For the fourth trial, $\lambda_T = 1.56\frac{k}{m}$ is used

$$\lambda_{T,4} = 1.56\frac{k}{m} - \frac{-m^3\left(1.56\frac{k}{m}\right)^3 + 5km^2\left(1.56\frac{k}{m}\right)^2 - 6k^2m\left(1.56\frac{k}{m}\right) + k^3}{-3m^3\left(1.56\frac{k}{m}\right)^2 + 10km^2\left(1.56\frac{k}{m}\right) - 6k^2m}$$

$$= 1.56\frac{k}{m} - \frac{-3.796k^3 + 12.17k^3 - 9.36k^3 + k^3}{-7.3k^2m + 15.6k^2m - 6k^2m}$$

$$= 1.56\frac{k}{m} - \frac{0.014k^3}{2.3k^2m} = 1.55\frac{k}{m}$$

Convergence achieved and one of the roots is $1.55\frac{k}{m}$.

Therefore, Eq. (6.28) can be simplified by factorizing it with the term $\left(\lambda - 1.55\frac{k}{m}\right)$:

$$\left(\lambda - 1.55\frac{k}{m}\right)\left(-m^3\lambda^2 + 3.45km^2\lambda - 0.6525k^2m\right) = 0$$

To solve the quadratic equation, $\lambda_{2,3} = \frac{-b\pm\sqrt{b^2-4ac}}{2a}$ is used, where $a = -m^3$, $b = 3.45km^2$, $c = -0.6525k^2m$

$$\lambda_{2,3} = \frac{-3.45km^2 \pm \sqrt{\left(3.45km^2\right)^2 - 4\left(-m^3\right)\left(-0.6525k^2m\right)}}{2\left(-m^3\right)}$$

$$= \frac{-3.45km^2 \pm \sqrt{11.9025k^2m^4 - 2.61k^2m^4}}{-2m^3}$$

$$= \frac{-3.45km^2 \pm 3.048km^2}{-2m^3}$$

By solving the cubic equation above for λ yields:

$$\lambda_2 = \frac{-3.45km^2 + 3.048km^2}{-2m^3} = 0.201\frac{k}{m}$$

$$\lambda_3 = \frac{-3.45km^2 - 3.048km^2}{-2m^3} = 3.25\frac{k}{m}$$

Therefore, the following can be concluded:

$$\text{For first mode, } \lambda_1 = \omega_{n,1}^2 = 0.201\frac{k}{m}$$

$$\text{For second mode, } \lambda_2 = \omega_{n,2}^2 = 1.55\frac{k}{m}$$

$$\text{For third mode, } \lambda_3 = \omega_{n,3}^2 = 3.25\frac{k}{m}$$

To obtain the first mode of vibration, substitute $\omega_{n,1}^2 = 0.201\frac{k}{m}$ into Eq. (6.26) after substituting the simplified form of $-\omega_n^2[m] + [k]$ as shown in (6.27):

$$\begin{bmatrix} -m\left(0.201\frac{k}{m}\right) + 2k & -k & 0 \\ -k & -m\left(0.201\frac{k}{m}\right) + 2k & -k \\ 0 & -k & -m\left(0.201\frac{k}{m}\right) + k \end{bmatrix} \begin{bmatrix} a_1 \\ a_2 \\ a_3 \end{bmatrix} = 0$$

Simplify the equation above yields:

$$\begin{bmatrix} 1.799k & -k & 0 \\ -k & 1.799k & -k \\ 0 & -k & 0.799k \end{bmatrix} \begin{bmatrix} a_1 \\ a_2 \\ a_3 \end{bmatrix} = 0$$

From the matrix above,

$$1.799ka_1 - ka_2 = 0 \tag{6.29}$$

$$-ka_1 + 1.799ka_2 - ka_3 = 0 \tag{6.30}$$

$$-ka_2 + 0.799ka_3 = 0 \tag{6.31}$$

From Eq. (6.29), express a_1 in term of a_2

$$a_1 = 0.556a_2 \tag{6.32}$$

From Eq. (6.31), express a_2 in term of a_3

$$a_2 = 0.799a_3 \tag{6.33}$$

By substituting Eq. (6.33) into (6.32) yields:

$$a_1 = 0.556(0.799a_3) = 0.444a_3$$

Therefore, for first mode of vibration,

$$\frac{a_2}{a_1} = \frac{1.799}{1}, \frac{a_3}{a_1} = \frac{2.252}{1}$$

To obtain the second mode of vibration, substitute $\omega_{n,2}^2 = 1.55\frac{k}{m}$ into Eq. (6.26) after substituting the simplified form of $-\omega_n^2[m] + [k]$ as shown in (6.27):

$$\begin{bmatrix} -m\left(1.55\frac{k}{m}\right) + 2k & -k & 0 \\ -k & -m\left(1.55\frac{k}{m}\right) + 2k & -k \\ 0 & -k & -m\left(1.55\frac{k}{m}\right) + k \end{bmatrix} \begin{bmatrix} a_1 \\ a_2 \\ a_3 \end{bmatrix} = 0$$

Simplify the equation above yields:

$$\begin{bmatrix} 0.45k & -k & 0 \\ -k & 0.45k & -k \\ 0 & -k & -0.55 \end{bmatrix} \begin{bmatrix} a_1 \\ a_2 \\ a_3 \end{bmatrix} = 0$$

From the matrix above,

$$0.45ka_1 - ka_2 = 0 \tag{6.34}$$

$$-ka_1 + 0.45ka_2 - ka_3 = 0 \tag{6.35}$$

$$-ka_2 - 0.55ka_3 = 0 \tag{6.36}$$

From Eq. (6.34), express a_1 in term of a_2

$$a_1 = 2.22a_2 \tag{6.37}$$

From Eq. (6.36), express a_2 in term of a_3

$$a_2 = -0.55a_3 \tag{6.38}$$

By substituting Eq. (6.38) into (6.37) (6.32) yields:

$$a_1 = 2.22(-0.55a_3) = -1.221a_3$$

Therefore, for second mode of vibration,

$$\frac{a_2}{a_1} = \frac{0.45}{1}, \frac{a_3}{a_1} = \frac{-0.819}{1}$$

To obtain the third mode of vibration, substitute $\omega_{n,3}^2 = 3.25\frac{k}{m}$ into Eq. (6.26) after substituting the simplified form of $-\omega_n^2[m] + [k]$ as shown in (6.27):

$$\begin{bmatrix} -m\left(3.25\frac{k}{m}\right)+2k & -k & 0 \\ -k & -m\left(3.25\frac{k}{m}\right)+2k & -k \\ 0 & -k & -m\left(3.25\frac{k}{m}\right)+k \end{bmatrix}\begin{bmatrix} a_1 \\ a_2 \\ a_3 \end{bmatrix} = 0$$

Simplify the equation above yields:

$$\begin{bmatrix} -1.25k & -k & 0 \\ -k & -1.25k & -k \\ 0 & -k & -2.25 \end{bmatrix}\begin{bmatrix} a_1 \\ a_2 \\ a_3 \end{bmatrix} = 0$$

From the matrix above,

$$-1.25ka_1 - ka_2 = 0 \tag{6.39}$$

$$-ka_1 - 1.25ka_2 - ka_3 = 0 \tag{6.40}$$

$$-ka_2 - 2.25ka_3 = 0 \tag{6.41}$$

From Eq. (6.39), express a_1 in term of a_2

$$a_1 = -0.8a_2 \tag{6.42}$$

From Eq. (6.41), express a_2 in term of a_3

$$a_2 = -2.25a_3 \tag{6.43}$$

By substituting Eq. (6.43) into (6.42) yields:

$$a_1 = -0.8(-2.25a_3) = 1.8a_3$$

Therefore, for third mode of vibration,

$$\frac{a_2}{a_1} = \frac{-1.25}{1}, \frac{a_3}{a_1} = \frac{0.556}{1}$$

Figure 6.8 shows the normal modes obtained from the above calculation:

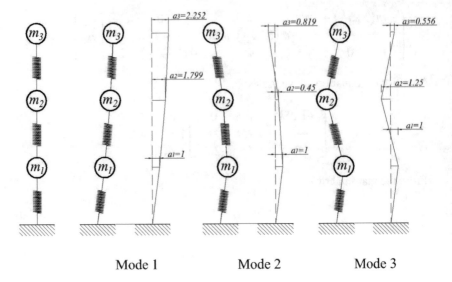

Mode 1   Mode 2   Mode 3

Fig. 6.8 Normal modes for structure in Example 6.1

6.5 Multi-degree of Freedom System Subjected to Impact Load

Considering an undamped 2-DOF system where each mass is subjected to different impact load as shown in Fig. 6.9.

The equation of motion for 2-DOF system can be derived based on the procedure outlined in 6.1 Equation of Motion. As a result, the equation of motion for the case illustrated above is:

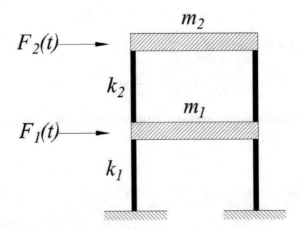

Fig. 6.9 2-DOF system subjected to impact load

$$\begin{bmatrix} m_1 & 0 \\ 0 & m_2 \end{bmatrix} \begin{bmatrix} \ddot{x}_1 \\ \ddot{x}_2 \end{bmatrix} + \begin{bmatrix} k_1 + k_2 & -k_2 \\ -k_2 & k_2 \end{bmatrix} \begin{bmatrix} x_1 \\ x_2 \end{bmatrix} = \begin{bmatrix} F_1(t) \\ F_2(t) \end{bmatrix}$$

The equation above can be simplified as:

$$[m]\{\ddot{x}(t)\} + [k]\{x(t)\} = \{f(t)\}$$

From Eq. (6.17), $x = \phi z$ and therefore $\ddot{x} = \phi \ddot{z}$, the equation above can be transformed to:

$$[m]\phi\{\ddot{z}(t)\} + [k]\phi\{z(t)\} = \{f(t)\}$$

By multiplying the terms on both sides with $[\phi]^T$ yields:

$$[\phi]^T[m]\phi\{\ddot{z}(t)\} + [\phi]^T[k]\phi\{z(t)\} = [\phi]^T\{f(t)\} \tag{6.44}$$

From the orthogonal nature of eigenvector, the following can be derived for mode-n:

$$[\phi_n]^T[m][\phi] = [M_n], \ [\phi_n]^T[m][\phi] = 1$$
$$[\phi_n]^T[k][\phi] = [K_n], \ [\phi_n]^T[k][\phi] = 1 \tag{6.45}$$

By substituting $[\phi_n]^T[m][\phi] = [M_n]$ and $[\phi_n]^T[k][\phi] = [K_n]$ into Eq. (6.44) yields:

$$[M_n]\{\ddot{z}(t)\} + [K_n]\{z(t)\} = [\phi]^T\{f(t)\}$$

By dividing the terms on both sides with $[M_n]$ yields:

$$\{\ddot{z}(t)\} + \frac{[K_n]}{[M_n]}\{z(t)\} = \frac{1}{[M_n]}[\phi]^T\{f(t)\}$$

By substituting Eq. (3.14) into equation above yields:

$$\{\ddot{z}(t)\} + \omega_{n,n}^2\{z(t)\} = \frac{1}{[M_n]}[\phi]^T\{f(t)\}$$

Based on the relationship arise from the orthogonality of eigenvector as shown in Eq. (6.45), equation above can be transformed to the following by substituting $M_n = 1$:

$$\{\ddot{z}(t)\} + \omega_{n,n}^2\{z(t)\} = [\phi]^T\{f(t)\}$$

By substituting Eq. (6.19) into equation:

$$\{\ddot{z}(t)\} + \omega_{n,n}^2\{z(t)\} = \{P(t)\}$$

Also, when the vibration modes are normalized, the component ϕ is defined as follows:

$$\phi = \begin{bmatrix} \phi_{11} & \phi_{12} \\ \phi_{21} & \phi_{22} \end{bmatrix}$$

The transposition of ϕ matrix yields:

$$\phi^T = \begin{bmatrix} \phi_{11} & \phi_{21} \\ \phi_{12} & \phi_{22} \end{bmatrix}$$

In complete matrix form, Eq. (6.19) can be written as:

$$\begin{bmatrix} P_1(t) \\ P_2(t) \end{bmatrix} = \begin{bmatrix} \phi_{11} & \phi_{21} \\ \phi_{12} & \phi_{22} \end{bmatrix} \begin{bmatrix} F_1(t) \\ F_2(t) \end{bmatrix} \tag{6.46}$$

From (6.46), the following equations can be produced:

$$P_1(t) = \phi_{11} F_1(t) + \phi_{21} F_2(t) \tag{6.47}$$

$$P_2(t) = \phi_{12} F_1(t) + \phi_{22} F_2(t) \tag{6.48}$$

When subjected to impact load, the stiffness component of structure plays a crucial role in defining the structural response. Therefore, the static modal displacement for first mode of vibration is:

$$z_{1,st} = \frac{P_{1,max}}{K_1}$$

Since $M_1 = 1$ as resulted from the orthogonality of eigenvectors, equation above can be transformed to:

$$z_{1,st} = \frac{P_{1,max} M_1}{K_1}$$

By substituting $\omega_{n,1} = \sqrt{\frac{K_n}{M_n}}$ into equation above yields:

$$z_{1,max} = \frac{P_{1,max}}{\omega_{n,1}^2}, \quad z_{1,max} = z_{1,st} \times DMF \tag{6.49}$$

Similarly, for second mode of vibration,

$$z_{2,\text{st}} = \frac{P_{2,\text{max}}}{\omega_{n,2}^2}, \quad z_{2,\text{max}} = z_{2,\text{st}} \times \text{DMF} \tag{6.50}$$

To obtain the actual response, Eq. (6.17) is required. There are two possible solutions for actual displacement by using Eqs. (6.49) and (6.50):

First solution:

$$x_{1,\text{max}} = |\phi_{11} z_{1,\text{max}}| + |\phi_{12} z_{2,\text{max}}|$$
$$x_{2,\text{max}} = |\phi_{21} z_{1,\text{max}}| + |\phi_{22} z_{2,\text{max}}| \tag{6.51}$$

Second solution:

$$x_{1,\text{max}} = \sqrt{(\phi_{11} z_{1,\text{max}})^2 + (\phi_{12} z_{2,\text{max}})^2}$$
$$x_{2,\text{max}} = \sqrt{(\phi_{21} z_{1,\text{max}})^2 + (\phi_{22} z_{2,\text{max}})^2} \tag{6.52}$$

The shear force developed in column (stiffness component) for storey i and mode of vibration j can be calculated using the following equation:

$$V_{ij} = z_{j,\text{max}}(\phi_{ij} - \phi_{(i-i),j})k_i \tag{6.53}$$

The shear force developed in first and second storey of structure as shown in Fig. 6.9 can be determined using equation below:

$$V_1 = \sqrt{V_{11}^2 + V_{12}^2}$$
$$V_2 = \sqrt{V_{21}^2 + V_{22}^2}$$

In generalized form, the shear force developed in storey i is:

$$V_i = \sqrt{\sum_{j=1}^{n} V_{ij}^2} \tag{6.54}$$

Example 6.2 Multi-Degree of Freedom System under Impact Load

Determine the shear force for each storey of structure under impact load, as shown in Fig. 6.10. The modulus of elasticity for steel should be taken as 200 GPa.

Solution

By referring to Table A.1 in Appendix, second moment of area, I for rectangular hollow section can be determined using the following equation:

$$I = \frac{bh^3}{12} - \frac{(b - 2t_w)(h - 2t_f)^3}{12} = \frac{0.1^4 - 0.08^4}{12} = 4.92 \times 10^{-6} \text{ m}^4$$

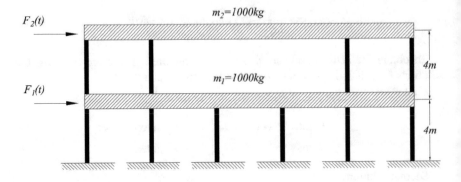

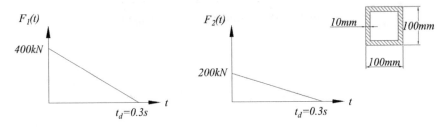

Fig. 6.10 Example 6.2

By referring to Table A.2 in Appendix, stiffness, k for fixed ends columns can be determined using the following equation:

$$k = \frac{12EI}{L^3} = \frac{2 \times 12 \times 200 \times 10^9 \times 4.92 \times 10^{-6}}{4^3} = 184,500 \text{ N/m}$$

For the first storey, six columns present and therefore the stiffness component is:

$$k_1 = 6k = 6 \times 184,500 = 1,107,000 \text{ N/m}$$

For second storey, four columns present and therefore the stiffness component is:

$$k_2 = 4k = 4 \times 184,500 = 738,000 \text{ N/m}$$

The equation of motion for this structure under free vibration is written by substituting the parameters into Eq. (6.15) to determine the natural frequency of each degree of freedom:

$$\begin{bmatrix} 1107000 + 738000 - 1000\omega_n^2 & -738000 \\ -738000 & 738000 - 1000\omega_n^2 \end{bmatrix} \begin{bmatrix} a_1 \\ a_2 \end{bmatrix} = \begin{bmatrix} 0 \\ 0 \end{bmatrix}$$

$$\begin{bmatrix} 1845000 - 1000\omega_n^2 & -738000 \\ -738000 & 738000 - 1000\omega_n^2 \end{bmatrix}\begin{bmatrix} a_1 \\ a_2 \end{bmatrix} = \begin{bmatrix} 0 \\ 0 \end{bmatrix} \qquad (6.55)$$

Since a_1 and a_2 are the solution to the matrix and they cannot be zero,

$$\begin{bmatrix} 1845000 - 1000\omega_n^2 & -738000 \\ -738000 & 738000 - 1000\omega_n^2 \end{bmatrix} = \begin{bmatrix} 0 \\ 0 \end{bmatrix}$$

The determinant of the matrix above must be zero to make the equation above valid:

$$\left(1845000 - 1000\omega_n^2\right)\left(738000 - 1000\omega_n^2\right) - (-738000)(-738000) = 0$$

Expand the equation above yields:

$$1000000\omega_n^4 + 2583 \times 10^6\omega_n^2 + 1.36161 \times 10^{12} - 5.44644 \times 10^{11} = 0$$
$$1000000\omega_n^4 + 2583 \times 10^6\omega_n^2 + 8.16966 \times 10^{11} = 0$$

Let $\lambda = \omega_n^2$ and subsequently solve the quadratic equation yields:

$$1000000\lambda^2 + 2583 \times 10^6\lambda + 8.16966 \times 10^{11} = 0$$

$$\lambda_{1,2} = \frac{2583 \times 10^6 \pm \sqrt{(-2583 \times 10^6)^2 - 4(1000000)\left(8.16966 \times 10^{11}\right)}}{2(1000000)}$$
$$= \frac{2583 \times 10^6 \pm 1845 \times 10^6}{2 \times 10^6}$$

From the expression above, the roots of the quadratic equation are:

$$\lambda_1 = \omega_{n,1}^2 = \frac{2583 - 1845}{2} = 369$$
$$\omega_{n,1} = 19.2 \text{ rad/s for first mode of vibration}$$
$$\lambda_2 = \omega_{n,2}^2 = \frac{2583 + 1845}{2} = 2214$$
$$\omega_{n,2} = 47.1 \text{ rad/s for second mode of vibration}$$

By substituting $\omega_{n,1}^2 = 369$ into Eq. (6.55) to obtain the amplitude of displacements for first mode of vibration:

$$\begin{bmatrix} 1845000 - 1000(369) & -738000 \\ -738000 & 738000 - 1000(369) \end{bmatrix}\begin{bmatrix} a_1 \\ a_2 \end{bmatrix} = \begin{bmatrix} 0 \\ 0 \end{bmatrix}$$

$$\begin{bmatrix} 1476000 & -738000 \\ -738000 & 369000 \end{bmatrix} \begin{bmatrix} a_1 \\ a_2 \end{bmatrix} = \begin{bmatrix} 0 \\ 0 \end{bmatrix}$$

From matrix above,

$$1476000a_1 - 738000a_2 = 0$$

$$\frac{a_{11}}{a_{21}} = \frac{1}{2}$$

By substituting $\omega_{n,2}^2 = 2214$ into Eq. (6.55) to obtain the amplitude of displacements for second mode of vibration:

$$\begin{bmatrix} 1845000 - 1000(2214) & -738000 \\ -738000 & 738000 - 1000(2214) \end{bmatrix} \begin{bmatrix} a_1 \\ a_2 \end{bmatrix} = \begin{bmatrix} 0 \\ 0 \end{bmatrix}$$

$$\begin{bmatrix} -369000 & -738000 \\ -738000 & -1476000 \end{bmatrix} \begin{bmatrix} a_1 \\ a_2 \end{bmatrix} = \begin{bmatrix} 0 \\ 0 \end{bmatrix}$$

From matrix above,

$$-369000a_1 - 738000a_2 = 0$$

$$\frac{a_{12}}{a_{22}} = -\frac{2}{1} = \frac{1}{-0.5}$$

The a component can be written in matrix form as:

$$[a] = \begin{bmatrix} a_{11} & a_{12} \\ a_{21} & a_{22} \end{bmatrix} = \begin{bmatrix} 1 & 1 \\ 2 & -0.5 \end{bmatrix}$$

To normalize the vibration, the relationship as per Eq. (6.16) is used:

$$\phi_{11} = \frac{a_{11}}{\sqrt{m_1 a_{11}^2 + m_2 a_{21}^2}} = \frac{1}{\sqrt{1000 \times (1)^2 + 1000 \times (2)^2}} = \frac{0.447}{\sqrt{1000}}$$

$$\phi_{21} = \frac{a_{21}}{\sqrt{m_1 a_{11}^2 + m_2 a_{21}^2}} = \frac{2}{\sqrt{1000 \times (1)^2 + 1000 \times (2)^2}} = \frac{0.894}{\sqrt{1000}}$$

$$\phi_{12} = \frac{a_{12}}{\sqrt{m_1 a_{12}^2 + m_2 a_{22}^2}} = \frac{1}{\sqrt{1000 \times (1)^2 + 1000 \times (-0.5)^2}} = \frac{0.894}{\sqrt{1000}}$$

$$\phi_{22} = \frac{a_{22}}{\sqrt{m_1 a_{12}^2 + m_2 a_{22}^2}} = \frac{-0.5}{\sqrt{1000 \times (1)^2 + 1000 \times (-0.5)^2}} = \frac{-0.447}{\sqrt{1000}}$$

Therefore,

$$\phi = \begin{bmatrix} \phi_{11} & \phi_{12} \\ \phi_{21} & \phi_{22} \end{bmatrix} = \frac{1}{\sqrt{1000}} \begin{bmatrix} 0.447 & 0.894 \\ 0.894 & -0.447 \end{bmatrix} = \begin{bmatrix} 0.0141 & 0.0283 \\ 0.0283 & -0.0141 \end{bmatrix}$$

From Eq. (6.46), $[P]$ can be determined

$$\begin{bmatrix} P_1 \\ P_2 \end{bmatrix} = \begin{bmatrix} 0.0141 & 0.0283 \\ 0.0283 & -0.0141 \end{bmatrix} \begin{bmatrix} 400000 \\ 200000 \end{bmatrix} = \begin{bmatrix} 11380 \\ 8560 \end{bmatrix}$$

The displacements of structure under first and second mode of vibration are:

$$z_1 = \frac{P_1}{\omega_{n,1}^2} = \frac{11380}{369} = 30.84 \, \text{m}$$

$$z_2 = \frac{P_2}{\omega_{n,2}^2} = \frac{8560}{2214} = 3.87 \, \text{m}$$

To calculate the displacement of structure under impact load, DMF is required. To obtain DMF from chart as shown in Fig. 5.17, the following calculation is necessary:

$$\frac{t_d}{T_1} = \frac{t_d}{\left(\frac{2\pi}{\omega_{n,1}} \right)} = \frac{0.3}{\left(\frac{2\pi}{19.2} \right)} = 0.917$$

$$\text{DMF} = 1.5$$

$$\frac{t_d}{T_2} = \frac{t_d}{\left(\frac{2\pi}{\omega_{n,2}} \right)} = \frac{0.3}{\left(\frac{2\pi}{47.1} \right)} = 2.25$$

$$\text{DMF} = 1.78$$

$$z_{1,\text{max}} = 30.84 \times 1.5 = 46.3 \, \text{m}$$

$$z_{2,\text{max}} = 3.87 \times 1.78 = 6.89 \, \text{m}$$

Based on first solution as shown in Eq. (6.51), the structural response for first and second vibration modes are:

$$x_{1,\text{max}} = |z_{1,\text{max}} \phi_{11}| + |z_{2,\text{max}} \phi_{12}| = |46.3 \times 0.0141| + |6.89 \times 0.0283|$$
$$= 0.85 \, \text{m}$$

$$x_{2,\text{max}} = |z_{1,\text{max}} \phi_{21}| + |z_{2,\text{max}} \phi_{22}| = |46.3 \times 0.0283| + |6.89 \times 0.0141|$$
$$= 1.41 \, \text{m}$$

Based on second solution as shown in Eq. (6.52), the structural response for first and second vibration modes are:

$$x_{1,max} = \sqrt{\left(z_{1,max}\phi_{11}\right)^2 + \left(z_{2,max}\phi_{12}\right)^2}$$
$$= \sqrt{(46.3 \times 0.0141)^2 + (6.89 \times 0.0283)^2} = 0.68 \, m$$
$$x_{2,max} = \sqrt{\left(z_{1,max}\phi_{21}\right)^2 + \left(z_{2,max}\phi_{22}\right)^2}$$
$$= \sqrt{(46.3 \times 0.0283)^2 + (6.89 \times -0.0141)^2} = 1.31 \, m$$

The shear force developed in columns at first storey can be determined using Eq. (6.54), where V_{11} and V_{12} are calculated using Eq. (6.53):

$$V_{11} = z_{1,max}(\phi_{11} - \phi_{01})k_1 = 46.3 \times 0.0141 \times 1107000 = 722.7 \, kN$$
$$V_{12} = z_{2,max}(\phi_{12} - \phi_{02})k_1 = 6.89 \times 0.0283 \times 1107000 = 215.9 \, kN$$
$$V_1 = \sqrt{722.7^2 + 215.9^2} = 754.3 \, kN$$

The shear force developed in columns at second storey can be determined using Eq. (6.54), where V_{11} and V_{12} are calculated using Eq. (6.53):

$$V_{21} = z_{1,max}(\phi_{21} - \phi_{11})k_2 = 46.3 \times (0.0283 - 0.0141) \times 738000 = 485.2 \, kN$$
$$V_{22} = z_{2,max}(\phi_{22} - \phi_{12})k_2 = 6.89 \times (-0.0141 - 0.0283) \times 738000 = 215.6 \, kN$$
$$V_2 = \sqrt{485.2^2 + 215.6^2} = 530.9 \, kN$$

6.6 Multi-degree of Freedom System Subjected to Earthquake Excitation

Based on Newton's second law of motion, external force due to earthquake excitation is defined as:

$$F_e(t) = m\ddot{y}(t)$$

Since the effect of the moment of inertia force due to earthquake excitation is opposite to the structure movement, the negative sign is needed when the force is inserted to the right-hand side of the equation of motion. Therefore, the equation for multi-degree of freedom system subjected to earthquake excitation can be derived as:

$$[M]\{\ddot{z}(t)\} + [C]\{\dot{z}(t)\} + [K]\{z(t)\} = -[M]\{\ddot{y}(t)\} \tag{6.56}$$

In this case, the earthquake excitation has been considered for force damped multi-degree of freedom system as idealized in Fig. 6.11.

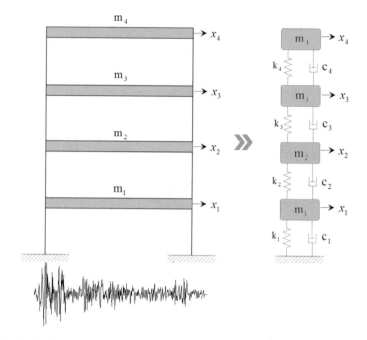

Fig. 6.11 Single-bay model frame as a representation for multi-degree of freedom system

$z(t)$, $\dot{z}(t)$ and $\ddot{z}(t)$ are actual responses of structure under earthquake excitation. To normalize these responses, they are usually being scaled down with scale factor ϕ. These results in $\psi(t)$, $\dot{\psi}(t)$ and $\ddot{\psi}(t)$, denotes the modal displacement, velocity and acceleration response of structure respectively. Inversely, by multiplying the factor ϕ to modal response, the actual response can be obtained:

$$z(t) = [\phi]\{\psi(t)\}, \dot{z}(t) = [\phi]\{\dot{\psi}(t)\}, \ddot{z}(t) = [\phi]\{\ddot{\psi}(t)\} \qquad (6.57)$$

By substituting Eq. (6.57) into Eq. (6.56) yields:

$$[M][\phi]\{\ddot{\psi}(t)\} + [C][\phi]\{\dot{\psi}(t)\} + [K][\phi]\{\psi(t)\} = -[M]\{\ddot{y}(t)\}$$

Multiply both sides with the term $[\phi_r]^T$ yields:

$$[\phi_r]^T[M][\phi]\{\ddot{\psi}(t)\} + [\phi_r]^T[C][\phi]\{\dot{\psi}(t)\} + [\phi_r]^T[K][\phi]\{\psi(t)\} = -[\phi_r]^T[M]\{\ddot{y}(t)\}$$

From the orthogonality properties,

$$[\phi_r]^T[M][\phi] = 0 \quad r \neq 0$$
$$[\phi_r]^T[C][\phi] = 0 \quad r \neq 0$$
$$[\phi_r]^T[K][\phi] = 0 \quad r \neq 0$$

It can be also considered for mode n:

$$M_n = [\phi_n]^T[M][\phi]$$
$$C_n = [\phi_n]^T[C][\phi]$$
$$K_n = [\phi_n]^T[K][\phi]$$
$$F_n = [\phi_n]^T[M]\{\ddot{y}(t)\} \tag{6.58}$$

Therefore, the equation of motion is:

$$[M_n]\{\ddot{\psi}\} + [C_n]\{\dot{\psi}\} + [K_n]\{\psi\} = -[\phi_r]^T[M]\{\ddot{y}(t)\}$$

Divide the terms on both sides by $[M_n]$ yields:

$$\{\ddot{\psi}\} + \frac{[C_n]}{[M_n]}\{\dot{\psi}\} + \frac{[K_n]}{[M_n]}\{\psi\} = -\frac{[\phi_r]^T[M]}{[M_n]}\{\ddot{y}(t)\}$$

By substituting Eqs. (3.14), (3.19) and Eq. (6.58) into equation above yields:

$$\{\ddot{\psi}\} + 2\xi\omega_n\{\dot{\psi}\} + \omega_n^2\{\psi\} = -\frac{[\phi_r]^T[M]}{[\phi_n]^T[M][\phi]}\{\ddot{y}(t)\}$$

In matrix form,

$$\begin{bmatrix} \ddot{\psi}_1 \\ \ddot{\psi}_2 \\ \vdots \\ \ddot{\psi}_n \end{bmatrix} + \begin{bmatrix} 2\xi\omega_1 & 0 & 0 & 0 \\ 0 & 2\xi\omega_2 & 0 & 0 \\ \vdots & \cdots & \cdots & \vdots \\ \cdots & \cdots & \cdots & 2\xi\omega_n \end{bmatrix} \begin{bmatrix} \dot{\psi}_1 \\ \dot{\psi}_2 \\ \vdots \\ \dot{\psi}_n \end{bmatrix} + \begin{bmatrix} \omega_1^2 & 0 & 0 & 0 \\ 0 & \omega_2^2 & 0 & 0 \\ \vdots & \cdots & \cdots & \vdots \\ \cdots & \cdots & \cdots & \omega_n^2 \end{bmatrix} \begin{bmatrix} \psi_1 \\ \psi_2 \\ \vdots \\ \psi_n \end{bmatrix} = \frac{1}{[M_n]} \begin{bmatrix} f_1(t) \\ f_2(t) \\ \vdots \\ f_n(t) \end{bmatrix}$$

For the sake of convenience, design usually conducted for the critical situation that structure is undertaken in term of maximum moment, shear force and displacement. In this case, maximum displacement is a significant factor in the design process against earthquake excitation. The following equations are applied during design stage:

$$Z = [\phi]\psi \text{ and } \psi = C_r S_{dr}$$

Therefore, the structural response under earthquake excitation can be expressed as:

$$Z = [\phi][C_r] \times S_{dr}$$

S_{dr} is to be obtained from the spectra chart, such as the one as shown in Fig. 5.20. The load, Q can be determined using:

$$Q_{ir} = m_i \omega_i^2 z_{ir}$$

Base shear force is defined as:

$$\sum_{i=1}^{n} Q_{ir}$$

Example 6.3 Multi-Degree of Freedom System under Earthquake Excitation
   Find the base shear of two-storey of steel frame system shown in Fig. 6.12 when
it's subjected to earthquake excitation noting that $\xi = 2\%$.

Solution

By referring to Table A.1 in Appendix, second moment of area, I for circular hollow
section can be determined using the following equation:

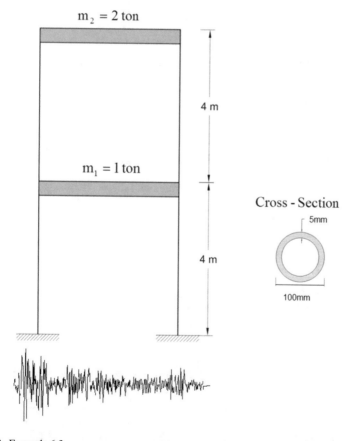

Fig. 6.12 Example 6.3

$$I = \frac{\pi d^4}{64} - \frac{\pi (d - 2t)^4}{64} = \frac{\pi (0.1)^4}{64} - \frac{\pi (0.1 - 2 \times 0.05)^4}{64} = 1.688 \times 10^{-6} \, \text{m}^4$$

By referring to Table A.2 in Appendix, stiffness, k for fixed ends column can be determined using the following equation:

$$k_1 = k_2 = 2 \times \frac{12EI}{L^3} = \frac{2 \times 12 \times 200 \times 10^9 \times 1.688 \times 10^{-6}}{4^3} = 126,600 \, \text{N/m}$$

For two-degree of freedom system, the following matrix is applied:

$$\begin{bmatrix} k_1 + k_2 - m_1\omega_n^2 & -k_2 \\ -k_2 & k_2 - m_2\omega_n^2 \end{bmatrix} \begin{bmatrix} a_1 \\ a_2 \end{bmatrix} = \begin{bmatrix} 0 \\ 0 \end{bmatrix}$$

By substituting all the parameters into the matrix above, and solve it yields:

$$10^4 \begin{bmatrix} 25.32 - 0.1\omega_n^2 & -12.26 \\ -12.26 & 12.26 - 0.2\omega_n^2 \end{bmatrix} \begin{bmatrix} a_1 \\ a_2 \end{bmatrix} = \begin{bmatrix} 0 \\ 0 \end{bmatrix}$$

$$(25.32 - 0.1\omega_n^2)(12.26 - 0.2\omega_n^2) - 150.31 = 0$$

$$312.96 - 5.064\omega_n^2 - 1.226\omega_n^2 + 0.02\omega_n^4 - 150.31 = 0$$

$$0.02\omega_n^4 - 6.29\omega_n^2 + 162.65 = 0$$

$$\omega_n^4 - 314.5\omega_n^2 + 8132.5 = 0$$

By solving the equation above yields:
$\omega_1^2 = 28.4, \omega_1 = \sqrt{28.4} = 5.33 \, \text{rad/s}$ and $T_1 = \frac{2\pi}{\omega_1} = \frac{2\pi}{5.33} = 1.18 \, \text{s}$ for first mode of vibration.
$\omega_2^2 = 286.1, \omega_2 = \sqrt{286.1} = 16.91 \, \text{rad/s}$ and $T_2 = \frac{2\pi}{\omega_2} = \frac{2\pi}{16.91} = 0.37 \, \text{s}$ for second mode of vibration.

For first mode of vibration:

$$\begin{bmatrix} 25.32 - 0.1(28.4) & -12.26 \\ -12.26 & 12.26 - 0.2(28.4) \end{bmatrix} \begin{bmatrix} a_{11} \\ a_{21} \end{bmatrix} = \begin{bmatrix} 0 \\ 0 \end{bmatrix}$$

$$\begin{bmatrix} 22.48 & -12.26 \\ -12.26 & 6.58 \end{bmatrix} \begin{bmatrix} a_{11} \\ a_{21} \end{bmatrix} = \begin{bmatrix} 0 \\ 0 \end{bmatrix}$$

$$22.48a_{11} - 12.26a_{21} = 0$$

$$\frac{a_{11}}{a_{21}} = \frac{1}{1.83}$$

For second mode of vibration:

$$\begin{bmatrix} 25.32 - 0.1(286.1) & -12.26 \\ -12.26 & 12.26 - 0.2(286.1) \end{bmatrix} \begin{bmatrix} a_{12} \\ a_{22} \end{bmatrix} = \begin{bmatrix} 0 \\ 0 \end{bmatrix}$$

$$\begin{bmatrix} -3.29 & -12.26 \\ -12.26 & -44.96 \end{bmatrix} \begin{bmatrix} a_{12} \\ a_{22} \end{bmatrix} = \begin{bmatrix} 0 \\ 0 \end{bmatrix}$$

$$-3.29 a_{12} - 12.26 a_{22} = 0$$

$$\frac{a_{12}}{a_{22}} = \frac{1}{-0.27}$$

The normalized mode of vibration can be determined using:

$$\phi_{ij} = \frac{a_{ij}}{\sqrt{\sum_{k=1}^{n} m_k a_{ij}^2}}, \quad \text{where} \quad \sqrt{\sum_{k=1}^{n} m_k a_{ij}^2} = \sqrt{m_1 a_{11}^2 + m_2 a_{21}^2}$$

For first mode of vibration:

$$\sqrt{\sum_{k=1}^{n} m_k a_{ij}^2} = \sqrt{m_1 a_{11}^2 + m_2 a_{21}^2} = \sqrt{1000 \times 1 + 2000 \times 1.83^2} = 87.74$$

$$\phi_{11} = \frac{1}{87.74} = 0.0114, \ \phi_{21} = \frac{1.83}{87.74} = 0.209$$

For second mode of vibration:

$$\sqrt{\sum_{k=1}^{n} m_k a_{ij}^2} = \sqrt{m_1 a_{12}^2 + m_2 a_{22}^2} = \sqrt{1000 \times 1 + 2000 \times -0.27^2} = 33.85$$

$$\phi_{11} = \frac{1}{33.85} = 0.0295, \ \phi_{21} = \frac{-0.27}{33.85} = -0.01$$

Therefore,

$$\phi_{ij} = \begin{bmatrix} 0.0114 & 0.0295 \\ 0.209 & -0.01 \end{bmatrix}$$

The mass contribution for each mode can be determined using:

$$C_r = \frac{\sum m_1 \phi_{ir}}{\sum m_i \phi_{ir}^2}$$

For first mode of vibration:

$$C_1 = \frac{m_1 \phi_{11} + m_2 \phi_{21}}{m_1 \phi_{11}^2 + m_2 \phi_{21}^2} = \frac{1000 \times 0.0114 + 2000 \times 0.209}{1000 \times 0.0114^2 + 2000 \times 0.209^2} = 4.91$$

For second mode of vibration:

$$C_1 = \frac{m_1\phi_{12} + m_2\phi_{22}}{m_1\phi_{12}^2 + m_2\phi_{22}^2} = \frac{1000 \times 0.0295 + 2000 \times -0.01}{1000 \times 0.0295^2 + 2000 \times 0.0.01^2} = 8.876$$

Based on Fig. 5.20, the following design values obtained for seismic design:
For first mode of vibration, $T_1 = 1.18\,\mathrm{s}$

$$S_{dr1} = 0.22\,\mathrm{m}$$

For first mode of vibration, $T_1 = 0.37\,\mathrm{s}$

$$S_{dr2} = 0.048\,\mathrm{m}$$

By using $Z = [\phi][C_r] \times S_{dr}$:

$$Z_{11} = 0.0114 \times 4.91 \times 0.22 = 0.0123\,\mathrm{m}$$
$$Z_{21} = 0.209 \times 4.91 \times 0.22 = 0.23\,\mathrm{m}$$
$$Z_{12} = 0.0295 \times 8.9 \times 0.048 = 0.0126\,\mathrm{m}$$
$$Z_{22} = -0.01 \times 8.9 \times 0.048 = 0.004\,\mathrm{m}$$

By using $Q_{ir} = m_i\omega_i^2 z_{ir}$:

$$Q_{11} = 1000 \times 28.4 \times 0.0123 = 349.32\,\mathrm{N}$$
$$Q_{21} = 2000 \times 28.4 \times 0.23 = 37904\,\mathrm{N}$$
$$Q_{12} = 1000 \times 286.1 \times 0.0126 = 3604.86\,\mathrm{N}$$
$$Q_{22} = 2000 \times 286.1 \times -0.004 = -2288.8\,\mathrm{N}$$

The base shear can be determined with $\sum\limits_{i=1}^{n} Q_{ir}$

$$V_b = Q_{11} + Q_{21} = 349.32 + 37904 = 38253.32\,\mathrm{N} = 38.25\,\mathrm{kN}$$
$$V_b = Q_{12} + Q_{22} = 3604.86 - 2288.8 = 1316.1\,\mathrm{N} = 1.32\,\mathrm{kN}$$

Figure 6.13 shows the lateral displacement and shear force due to earthquake excitation for first and second mode of vibration.

6.7   Iterative Method

Iterative method trials the answer for normalized mode shapes until convergence is achieved. In addition to that, the corresponding natural frequency can be obtained. Consider a three-storey building idealized as shown in Fig. 6.14.

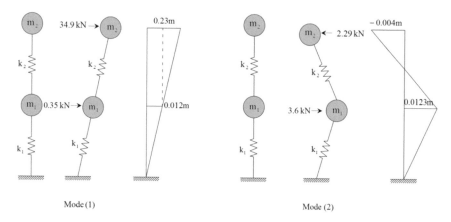

Mode (1)

Mode (2)

Fig. 6.13 Lateral displacement and shear force of Example 6.3 for each mode of vibration

Under undamped free vibration, the generated inertial force deploys the resistance of stiffness components and therefore,

$$m\ddot{x} = kx$$

By expressing the displacement response, x in terms of other parameters:

$$x = \frac{m}{k}\ddot{x}$$

Stiffness matrix, $[k]$ is corresponding to the stiffness component. When inversed, it is equivalent to flexibility matrix $[f]$, which defines the amount of force required to produce 1 unit of displacement. Therefore in matrix form, the equation above can be written as:

$$\{x\} = [f][m]\{\ddot{x}\}$$

When written in full form, the matrix above becomes:

$$\begin{Bmatrix} x_1 \\ x_2 \\ x_3 \end{Bmatrix} = \begin{bmatrix} f_{11} & f_{12} & f_{13} \\ f_{21} & f_{22} & f_{23} \\ f_{31} & f_{32} & f_{33} \end{bmatrix} \begin{bmatrix} m_1 & 0 & 0 \\ 0 & m_2 & 0 \\ 0 & 0 & m_3 \end{bmatrix} \begin{Bmatrix} \ddot{x}_1 \\ \ddot{x}_2 \\ \ddot{x}_3 \end{Bmatrix} \tag{6.59}$$

For harmonic vibration, the structural response can be expressed as follows, where P denotes the natural frequency for a specific mode:

$$x_1 = a_1 \sin Pt, \ \dot{x}_1 = a_1 P \cos Pt, \ \ddot{x}_1 = -a_1 P^2 \sin Pt$$
$$x_2 = a_2 \sin Pt, \ \dot{x}_2 = a_2 P \cos Pt, \ \ddot{x}_2 = -a_2 P^2 \sin Pt$$

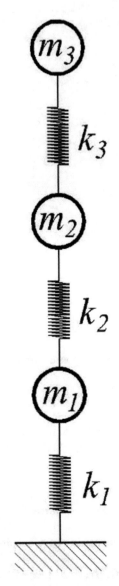

Fig. 6.14 Three-storey building idealized as a 3-DOF system

$$x_3 = a_3 \sin Pt, \, \dot{x}_3 = a_3 P \cos Pt, \, \ddot{x}_3 = -a_3 P^2 \sin Pt \qquad (6.60)$$

By substituting Eq. (6.60) into Eq. (6.59) yields the following:

$$\begin{Bmatrix} a_1 \sin Pt \\ a_2 \sin Pt \\ a_3 \sin Pt \end{Bmatrix} = \begin{bmatrix} f_{11} & f_{12} & f_{13} \\ f_{21} & f_{22} & f_{23} \\ f_{31} & f_{32} & f_{33} \end{bmatrix} \begin{bmatrix} m_1 & 0 & 0 \\ 0 & m_2 & 0 \\ 0 & 0 & m_3 \end{bmatrix} \begin{Bmatrix} -a_1 P^2 \sin Pt \\ -a_2 P^2 \sin Pt \\ -a_3 P^2 \sin Pt \end{Bmatrix}$$

By eliminating the common term $\sin Pt$ on both sides:

$$\begin{Bmatrix} a_1 \\ a_2 \\ a_3 \end{Bmatrix} = \begin{bmatrix} f_{11} & f_{12} & f_{13} \\ f_{21} & f_{22} & f_{23} \\ f_{31} & f_{32} & f_{33} \end{bmatrix} \begin{bmatrix} m_1 & 0 & 0 \\ 0 & m_2 & 0 \\ 0 & 0 & m_3 \end{bmatrix} \begin{Bmatrix} -a_1 P^2 \\ -a_2 P^2 \\ -a_3 P^2 \end{Bmatrix}$$

Simplify the terms on right-hand side by factorizing it with term P^2. The negative sign denotes the responses are to be in opposite directions to achieve equilibrium. In this case, only the magnitude of displacement is of interest. Therefore, by taking the absolute value:

$$\begin{Bmatrix} a_1 \\ a_2 \\ a_3 \end{Bmatrix} = P^2 \begin{bmatrix} f_{11} & f_{12} & f_{13} \\ f_{21} & f_{22} & f_{23} \\ f_{31} & f_{32} & f_{33} \end{bmatrix} \begin{bmatrix} m_1 & 0 & 0 \\ 0 & m_2 & 0 \\ 0 & 0 & m_3 \end{bmatrix} \begin{Bmatrix} a_1 \\ a_2 \\ a_3 \end{Bmatrix}$$

Bring over the term P^2 to the opposite side yields:

$$\frac{1}{P^2} \begin{Bmatrix} a_1 \\ a_2 \\ a_3 \end{Bmatrix} = \begin{bmatrix} f_{11} & f_{12} & f_{13} \\ f_{21} & f_{22} & f_{23} \\ f_{31} & f_{32} & f_{33} \end{bmatrix} \begin{bmatrix} m_1 & 0 & 0 \\ 0 & m_2 & 0 \\ 0 & 0 & m_3 \end{bmatrix} \begin{Bmatrix} a_1 \\ a_2 \\ a_3 \end{Bmatrix} \qquad (6.61)$$

In terms of absolute value, the value for flexibility is always less than that for stiffness because:

$$f = k^{-1} \qquad (6.62)$$

Therefore, when flexibility matrix is used, the resultant displacement is the largest among all the possible modes of vibration. Thus, the structural response for the first mode of vibration can be determined. By writing the Eq. (6.61) in matrix form,

$$\frac{1}{P^2} \begin{Bmatrix} a_1 \\ a_2 \\ a_3 \end{Bmatrix} = [f][m] \begin{Bmatrix} a_1 \\ a_2 \\ a_3 \end{Bmatrix}$$

When the displacement is normalized, the matrix above can simply be written as follow for a specific mode of vibration, say first mode:

$$\frac{1}{P^2} \begin{Bmatrix} \phi_{11} \\ \phi_{21} \\ \phi_{31} \end{Bmatrix} = [f][m] \begin{Bmatrix} \phi_{11} \\ \phi_{21} \\ \phi_{31} \end{Bmatrix} \qquad (6.63)$$

It can also be written as follows if eigenvalue λ is introduced to the equation above:

$$\lambda \left\{ \begin{array}{c} \phi_{11} \\ \phi_{21} \\ \phi_{31} \end{array} \right\} = [f][m] \left\{ \begin{array}{c} \phi_{11} \\ \phi_{21} \\ \phi_{31} \end{array} \right\}$$

From equation above, the following can be obtained:

$$\frac{1}{P^2} = [f][m]$$

By taking the inverse of both sides yields:

$$P^2 = [m]^{-1}[f]^{-1}$$

By applying the relationship as stated in Eq. (6.62), the equation above can be transformed to:

$$P^2 = [m]^{-1}[k]$$

When written in the form similar to Eq. (6.63),

$$P^2 \left\{ \begin{array}{c} \phi_{11} \\ \phi_{21} \\ \phi_{31} \end{array} \right\} = [m]^{-1}[k] \left\{ \begin{array}{c} \phi_{11} \\ \phi_{21} \\ \phi_{31} \end{array} \right\} \qquad (6.64)$$

When stiffness matrix is used, the resultant displacement is the smallest among all the possible modes of vibration. By doing so, the structural response for the least critical mode, e.g. third mode of vibration for a 3-DOF system can be obtained.

The procedure to perform iterative method for first mode of vibration is as follows:

(1) Determine the mass matrix $[m]$ and flexibility matrix $[f]$.
(2) Substitute the matrices into Eq. (6.63).
(3) Arbitrarily set the value for $\{\phi\}$ for the first trial.
4)   Calculate the resultant value using right-hand side of Eq. (6.63), i.e. $[f][m]\{\phi\}$.
(5) Normalize the result with respect to any of the degree of freedom. This is done by taking the magnitude of that degree of freedom as scale factor, say α and divide all the resultant component with that value. Note that when a specific component is chosen as such reference, it shall be used as a reference until the end of the process.

$$\{\phi\} = \alpha\{\phi/\alpha\}$$

(6) Check the deviation between the normalized result and the trial answer from Step 3. If it is acceptable, then the result will be adopted. Otherwise, repeat Step 3 to 6 using the calculated result as subsequent trial.

(7) The natural frequency for first mode of vibration is defined as follows:

$$\lambda = \frac{1}{P^2} = \alpha$$

To determine the responses for the least critical mode, Step 1 should be proceeded with stiffness matrix instead of flexibility matrix. Equation (6.64) will be used instead of Eq. (6.63), and the natural frequency P can be determined simply by finding the square root of scale factor α.

Example 6.4 Application of iterative method

Solve for the fundamental frequency for structure in Fig. 6.15 using iterative method.

Solution

The mass and stiffness matrices of the structure are as follow:

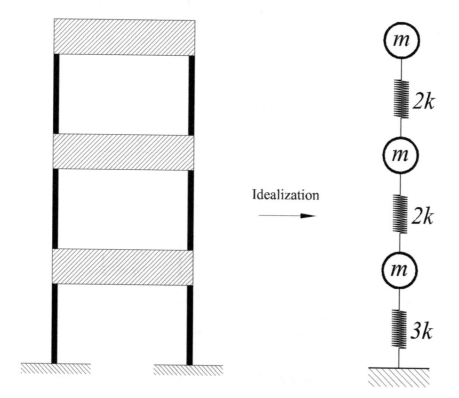

Fig. 6.15 3-DOF system for Example 6.4

$$[m] = \begin{bmatrix} m & 0 & 0 \\ 0 & m & 0 \\ 0 & 0 & m \end{bmatrix} = m \begin{bmatrix} 1 & 0 & 0 \\ 0 & 1 & 0 \\ 0 & 0 & 1 \end{bmatrix}$$

$$[k] = \begin{bmatrix} 3k+2k & -2k & 0 \\ -2k & 2k+2k & -2k \\ 0 & -2k & 2k \end{bmatrix} = \begin{bmatrix} 5k & -2k & 0 \\ -2k & 4k & -2k \\ 0 & -2k & 2k \end{bmatrix}$$

The flexibility matrix, $[f]$ is required to proceed with iterative method. It can be determined by finding the inverse of stiffness matrix $[k]$. The determinant of $[k]$ is as follows:

$$|k| = 5k[(4k)(2k) - (-2k)(-2k)] - (-2k)[(-2k)(2k)] + 0$$
$$= 5k(4k^2) + 2k(-4k^2) = 12k^3$$

Since $[k]$ is symmetrical, $[k]^T = [k]$. Next, determine the adjoint for matrix $[k]$:

$$Adj(k) = \begin{bmatrix} (4k)(2k)-(-2k)(-2k) & -(-2k)(2k) & (-2k)(-2k) \\ -(-2k)(2k) & (5k)(2k) & -(5k)(-2k) \\ (-2k)(-2k) & -(5k)(-2k) & (5k)(4k)-(-2k)(-2k) \end{bmatrix}$$

$$= \begin{bmatrix} 4k^2 & 4k^2 & 4k^2 \\ 4k^2 & 10k^2 & 10k^2 \\ 4k^2 & 10k^2 & 16k^2 \end{bmatrix}$$

The inverse of $[k]$ is defined as follows:

$$[f] = [k]^{-1} = \frac{1}{|k|} adj(k) = \frac{1}{12k^3} \begin{bmatrix} 4k^2 & 4k^2 & 4k^2 \\ 4k^2 & 10k^2 & 10k^2 \\ 4k^2 & 10k^2 & 16k^2 \end{bmatrix} = \frac{1}{6k} \begin{bmatrix} 2 & 2 & 2 \\ 2 & 5 & 5 \\ 2 & 5 & 8 \end{bmatrix}$$

By substituting the $[f]$ and $[m]$ into Eq. (6.63):

$$\frac{1}{6k} \begin{bmatrix} 2 & 2 & 2 \\ 2 & 5 & 5 \\ 2 & 5 & 8 \end{bmatrix} \times m \begin{bmatrix} 1 & 0 & 0 \\ 0 & 1 & 0 \\ 0 & 0 & 1 \end{bmatrix} \begin{Bmatrix} \phi_{11} \\ \phi_{21} \\ \phi_{31} \end{Bmatrix} = \frac{1}{P^2} \begin{Bmatrix} \phi_{11} \\ \phi_{21} \\ \phi_{31} \end{Bmatrix}$$

$$\frac{m}{6k} \begin{bmatrix} 2 & 2 & 2 \\ 2 & 5 & 5 \\ 2 & 5 & 8 \end{bmatrix} \begin{Bmatrix} \phi_{11} \\ \phi_{21} \\ \phi_{31} \end{Bmatrix} = \frac{1}{P^2} \begin{Bmatrix} \phi_{11} \\ \phi_{21} \\ \phi_{31} \end{Bmatrix} \qquad (6.65)$$

For the first iteration, let $\begin{Bmatrix} \phi_{11} \\ \phi_{21} \\ \phi_{31} \end{Bmatrix} = \begin{Bmatrix} 1 \\ 1 \\ 1 \end{Bmatrix}$ and by substituting it into Eq. (6.65):

$$\frac{m}{6k} \begin{bmatrix} 2 & 2 & 2 \\ 2 & 5 & 5 \\ 2 & 5 & 8 \end{bmatrix} \begin{Bmatrix} 1 \\ 1 \\ 1 \end{Bmatrix} = \begin{Bmatrix} 6 \\ 12 \\ 15 \end{Bmatrix}$$

Let the displacement at the top storey be the reference. In other words, after every iteration, the $\{\phi\}$ will be scaled by dividing the value obtained for ϕ_{13}. After normalization the above expression becomes:

$$\frac{m}{6k} \begin{bmatrix} 2 & 2 & 2 \\ 2 & 5 & 5 \\ 2 & 5 & 8 \end{bmatrix} \begin{Bmatrix} 1 \\ 1 \\ 1 \end{Bmatrix} = 15 \begin{Bmatrix} 0.4 \\ 0.8 \\ 1 \end{Bmatrix}$$

For the second iteration, let $\begin{Bmatrix} \phi_{11} \\ \phi_{21} \\ \phi_{31} \end{Bmatrix} = \begin{Bmatrix} 0.4 \\ 0.8 \\ 1 \end{Bmatrix}$ and by substituting it into Eq. (6.65):

$$\frac{m}{6k} \begin{bmatrix} 2 & 2 & 2 \\ 2 & 5 & 5 \\ 2 & 5 & 8 \end{bmatrix} \begin{Bmatrix} 0.4 \\ 0.8 \\ 1 \end{Bmatrix} = \begin{Bmatrix} 4.4 \\ 9.8 \\ 12.8 \end{Bmatrix}$$

After normalization, the above expression becomes:

$$\frac{m}{6k} \begin{bmatrix} 2 & 2 & 2 \\ 2 & 5 & 5 \\ 2 & 5 & 8 \end{bmatrix} \begin{Bmatrix} 0.4 \\ 0.8 \\ 1 \end{Bmatrix} = 12.8 \begin{Bmatrix} 0.343 \\ 0.766 \\ 1 \end{Bmatrix}$$

For the third iteration, let $\begin{Bmatrix} \phi_{11} \\ \phi_{21} \\ \phi_{31} \end{Bmatrix} = \begin{Bmatrix} 0.343 \\ 0.766 \\ 1 \end{Bmatrix}$ and by substituting it into Eq. (6.65):

$$\frac{m}{6k} \begin{bmatrix} 2 & 2 & 2 \\ 2 & 5 & 5 \\ 2 & 5 & 8 \end{bmatrix} \begin{Bmatrix} 0.343 \\ 0.766 \\ 1 \end{Bmatrix} = \begin{Bmatrix} 4.218 \\ 9.516 \\ 12.516 \end{Bmatrix}$$

After normalization, the above expression becomes:

$$\frac{m}{6k} \begin{bmatrix} 2 & 2 & 2 \\ 2 & 5 & 5 \\ 2 & 5 & 8 \end{bmatrix} \begin{Bmatrix} 0.343 \\ 0.766 \\ 1 \end{Bmatrix} = 12.516 \begin{Bmatrix} 0.337 \\ 0.760 \\ 1 \end{Bmatrix}$$

For the fourth iteration, let $\begin{Bmatrix} \phi_{11} \\ \phi_{21} \\ \phi_{31} \end{Bmatrix} = \begin{Bmatrix} 0.337 \\ 0.760 \\ 1 \end{Bmatrix}$ and by substituting it into Eq. (6.65):

$$\frac{m}{6k} \begin{bmatrix} 2\ 2\ 2 \\ 2\ 5\ 5 \\ 2\ 5\ 8 \end{bmatrix} \begin{Bmatrix} 0.337 \\ 0.760 \\ 1 \end{Bmatrix} = \begin{Bmatrix} 4.194 \\ 9.474 \\ 12.474 \end{Bmatrix}$$

After normalization, the above expression becomes:

$$\frac{m}{6k} \begin{bmatrix} 2\ 2\ 2 \\ 2\ 5\ 5 \\ 2\ 5\ 8 \end{bmatrix} \begin{Bmatrix} 0.337 \\ 0.760 \\ 1 \end{Bmatrix} = 12.474 \begin{Bmatrix} 0.336 \\ 0.759 \\ 1 \end{Bmatrix}$$

Convergence achieved and therefore, $\lambda = 12.474$.

$$\frac{1}{P^2} = \lambda = 12.474 \times \frac{m}{6k}$$

$$P^2 = \frac{1}{12.474} \times \frac{6k}{m} = 0.48 \frac{k}{m}$$

The fundamental frequency of structure is $0.693\sqrt{\frac{k}{m}}$.

Example 6.5 Normal modes determination using iterative method

Determine the most and least critical modes of vibration for an idealized structure as shown in Fig. 6.16.

Solution

The mass and stiffness matrices of the structure are as follow:

$$[m] = \begin{bmatrix} m\ 0\ 0 \\ 0\ m\ 0 \\ 0\ 0\ m \end{bmatrix} = m \begin{bmatrix} 1\ 0\ 0 \\ 0\ 1\ 0 \\ 0\ 0\ 1 \end{bmatrix}$$

$$[k] = \begin{bmatrix} k+k & -k & 0 \\ -k & k+k & -k \\ 0 & -k & k \end{bmatrix} = \begin{bmatrix} 2k & -k & 0 \\ -k & 2k & -k \\ 0 & -k & k \end{bmatrix}$$

To determine the least critical mode of vibration, Eq. (6.64) is adopted:

$$[m]^{-1}[k]\{\phi\} = P^2\{\phi\}$$

By substituting the stiffness matrix into the equation above yields:

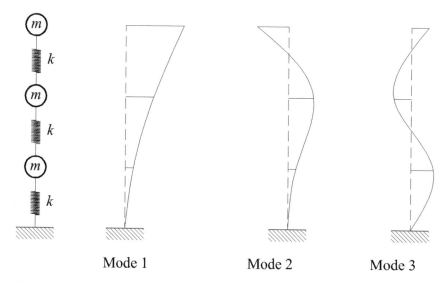

Fig. 6.16 Idealized structure and its anticipated modes of vibration for Example 6.5

$$\frac{1}{m}\begin{bmatrix} 1 & 0 & 0 \\ 0 & 1 & 0 \\ 0 & 0 & 1 \end{bmatrix}\begin{bmatrix} 2k & -k & 0 \\ -k & 2k & -k \\ 0 & -k & k \end{bmatrix}\begin{Bmatrix} \phi_{11} \\ \phi_{21} \\ \phi_{31} \end{Bmatrix} = P^2\begin{Bmatrix} \phi_{11} \\ \phi_{21} \\ \phi_{31} \end{Bmatrix}$$

$$\frac{k}{m}\begin{bmatrix} 2 & -1 & 0 \\ -1 & 2 & -1 \\ 0 & -1 & 1 \end{bmatrix}\begin{Bmatrix} \phi_{11} \\ \phi_{21} \\ \phi_{31} \end{Bmatrix} = P^2\begin{Bmatrix} \phi_{11} \\ \phi_{21} \\ \phi_{31} \end{Bmatrix} \qquad (6.66)$$

The third mode of vibration is the least critical as all adjacent masses are moving in opposite direction and this yields the least amount of displacement. By looking at the generalized mode shape, assuming $\begin{Bmatrix} \phi_{11} \\ \phi_{21} \\ \phi_{31} \end{Bmatrix} = \begin{Bmatrix} 1 \\ -1 \\ 1 \end{Bmatrix}$. By substituting the assumption into Eq. (6.66):

$$\frac{k}{m}\begin{bmatrix} 2 & -1 & 0 \\ -1 & 2 & -1 \\ 0 & -1 & 1 \end{bmatrix}\begin{Bmatrix} 1 \\ -1 \\ 1 \end{Bmatrix} = \begin{Bmatrix} 3 \\ -4 \\ 2 \end{Bmatrix}$$

Use the first storey as reference by dividing all terms in $\{\phi\}$ by ϕ_{11} after each iteration. After normalization, the above expression becomes:

$$\frac{k}{m}\begin{bmatrix} 2 & -1 & 0 \\ -1 & 2 & -1 \\ 0 & -1 & 1 \end{bmatrix}\begin{Bmatrix} 1 \\ -1 \\ 1 \end{Bmatrix} = 3\begin{Bmatrix} 1 \\ -1.33 \\ 0.66 \end{Bmatrix}$$

For the second iteration, substituting $\begin{Bmatrix} \phi_{11} \\ \phi_{21} \\ \phi_{31} \end{Bmatrix} = \begin{Bmatrix} 1 \\ -1.33 \\ 0.66 \end{Bmatrix}$ into Eq. (6.66):

$$\frac{k}{m} \begin{bmatrix} 2 & -1 & 0 \\ -1 & 2 & -1 \\ 0 & -1 & 1 \end{bmatrix} \begin{Bmatrix} 1 \\ -1.33 \\ 0.66 \end{Bmatrix} = \begin{Bmatrix} 3.33 \\ -4.33 \\ 2 \end{Bmatrix}$$

After normalization, the above expression becomes:

$$\frac{k}{m} \begin{bmatrix} 2 & -1 & 0 \\ -1 & 2 & -1 \\ 0 & -1 & 1 \end{bmatrix} \begin{Bmatrix} 1 \\ -1.33 \\ 0.66 \end{Bmatrix} = 3.33 \begin{Bmatrix} 1 \\ -1.3 \\ 0.6 \end{Bmatrix}$$

For the third iteration, substituting $\begin{Bmatrix} \phi_{11} \\ \phi_{21} \\ \phi_{31} \end{Bmatrix} = \begin{Bmatrix} 1 \\ -1.3 \\ 0.6 \end{Bmatrix}$ into Eq. (6.66):

$$\frac{k}{m} \begin{bmatrix} 2 & -1 & 0 \\ -1 & 2 & -1 \\ 0 & -1 & 1 \end{bmatrix} \begin{Bmatrix} 1 \\ -1.3 \\ 0.6 \end{Bmatrix} = \begin{Bmatrix} 3.3 \\ -4.2 \\ 1.9 \end{Bmatrix}$$

After normalization, the above expression becomes:

$$\frac{k}{m} \begin{bmatrix} 2 & -1 & 0 \\ -1 & 2 & -1 \\ 0 & -1 & 1 \end{bmatrix} \begin{Bmatrix} 1 \\ -1.3 \\ 0.6 \end{Bmatrix} = 3.3 \begin{Bmatrix} 1 \\ -1.27 \\ 0.58 \end{Bmatrix}$$

For the fourth iteration, substituting $\begin{Bmatrix} \phi_{11} \\ \phi_{21} \\ \phi_{31} \end{Bmatrix} = \begin{Bmatrix} 1 \\ -1.27 \\ 0.58 \end{Bmatrix}$ into Eq. (6.66):

$$\frac{k}{m} \begin{bmatrix} 2 & -1 & 0 \\ -1 & 2 & -1 \\ 0 & -1 & 1 \end{bmatrix} \begin{Bmatrix} 1 \\ -1.27 \\ 0.58 \end{Bmatrix} = \begin{Bmatrix} 3.27 \\ -4.12 \\ 1.85 \end{Bmatrix}$$

After normalization, the above expression becomes:

$$\frac{k}{m} \begin{bmatrix} 2 & -1 & 0 \\ -1 & 2 & -1 \\ 0 & -1 & 1 \end{bmatrix} \begin{Bmatrix} 1 \\ -1.27 \\ 0.58 \end{Bmatrix} = 3.27 \begin{Bmatrix} 1 \\ -1.26 \\ 0.56 \end{Bmatrix}$$

For the fifth iteration, substituting $\begin{Bmatrix} \phi_{11} \\ \phi_{21} \\ \phi_{31} \end{Bmatrix} = \begin{Bmatrix} 1 \\ -1.26 \\ 0.56 \end{Bmatrix}$ into Eq. (6.66):

$$\frac{k}{m} \begin{bmatrix} 2 & -1 & 0 \\ -1 & 2 & -1 \\ 0 & -1 & 1 \end{bmatrix} \begin{Bmatrix} 1 \\ -1.26 \\ 0.56 \end{Bmatrix} = \begin{Bmatrix} 3.26 \\ -4.08 \\ 1.82 \end{Bmatrix}$$

After normalization, the above expression becomes:

$$\frac{k}{m} \begin{bmatrix} 2 & -1 & 0 \\ -1 & 2 & -1 \\ 0 & -1 & 1 \end{bmatrix} \begin{Bmatrix} 1 \\ -1.26 \\ 0.56 \end{Bmatrix} = 3.26 \begin{Bmatrix} 1 \\ -1.25 \\ 0.56 \end{Bmatrix}$$

Convergence achieved and therefore, $P^2 = 3.26\frac{k}{m}$. The frequency for third mode of vibration is $1.806\sqrt{\frac{k}{m}}$.

To determine the most critical mode of vibration, Eq. (6.63) is adopted:

$$[f][m]\{\phi\} = \frac{1}{P^2}\{\phi\}$$

$$[k]^{-1}[m]\{\phi\} = \frac{1}{P^2}\{\phi\}$$

To find the flexibility matrix $[f]$, the determinant of stiffness matrix $[k]$ must be determined:

$$|k| = 2k[(2k)(k) - (-k)(-k)] - (-k)[(-k)(k)] + 0 = 2k(k^2) + k(-k^2)$$
$$= k^3$$

Since $[k]$ is symmetrical, $[k]^{\mathrm{T}} = [k]$. Next, determine the adjoint for matrix $[k]$:

$$\mathrm{Adj}(k) = \begin{bmatrix} (2k)(k) - (-k)(-k) & -(-k)(k) & (-k)(-k) \\ -(-k)(k) & (2k)(k) & -(2k)(-k) \\ (-k)(-k) & -(2k)(-k) & (2k)(2k) - (-k)(-k) \end{bmatrix}$$

$$= \begin{bmatrix} k^2 & k^2 & k^2 \\ k^2 & 2k^2 & 2k^2 \\ k^2 & 2k^2 & 3k^2 \end{bmatrix}$$

The flexibility matrix $[f]$ is defined as follows:

$$[f] = [k]^{-1} = \frac{1}{|k|}\text{adj}(k) = \frac{1}{k^3}\begin{bmatrix} k^2 & k^2 & k^2 \\ k^2 & 2k^2 & 2k^2 \\ k^2 & 2k^2 & 3k^2 \end{bmatrix} = \frac{1}{k}\begin{bmatrix} 1 & 1 & 1 \\ 1 & 2 & 2 \\ 1 & 2 & 3 \end{bmatrix}$$

By substituting the $[f]$ and $[m]$ into Eq. (6.63):

$$\frac{1}{k}\begin{bmatrix} 1 & 1 & 1 \\ 1 & 2 & 2 \\ 1 & 2 & 3 \end{bmatrix} \times m \begin{bmatrix} 1 & 0 & 0 \\ 0 & 1 & 0 \\ 0 & 0 & 1 \end{bmatrix} \begin{Bmatrix} \phi_{11} \\ \phi_{21} \\ \phi_{31} \end{Bmatrix} = \frac{1}{P^2}\begin{Bmatrix} \phi_{11} \\ \phi_{21} \\ \phi_{31} \end{Bmatrix}$$

$$\frac{m}{k}\begin{bmatrix} 1 & 1 & 1 \\ 1 & 2 & 2 \\ 1 & 2 & 3 \end{bmatrix} \begin{Bmatrix} \phi_{11} \\ \phi_{21} \\ \phi_{31} \end{Bmatrix} = \frac{1}{P^2}\begin{Bmatrix} \phi_{11} \\ \phi_{21} \\ \phi_{31} \end{Bmatrix} \qquad (6.67)$$

The first mode of vibration is the most critical as all masses are moving in the same direction and this yields the greatest amount of displacement. By looking at the generalized mode shape, assuming $\begin{Bmatrix} \phi_{11} \\ \phi_{21} \\ \phi_{31} \end{Bmatrix} = \begin{Bmatrix} 1 \\ 1 \\ 1 \end{Bmatrix}$. By substituting the assumption into Eq. (6.67):

$$\frac{m}{k}\begin{bmatrix} 1 & 1 & 1 \\ 1 & 2 & 2 \\ 1 & 2 & 3 \end{bmatrix} \begin{Bmatrix} 1 \\ 1 \\ 1 \end{Bmatrix} = \begin{Bmatrix} 3 \\ 5 \\ 6 \end{Bmatrix}$$

Use the first storey as reference by dividing all terms in $\{\phi\}$ by ϕ_{11} after each iteration. After normalization, the above expression becomes:

$$\frac{m}{k}\begin{bmatrix} 1 & 1 & 1 \\ 1 & 2 & 2 \\ 1 & 2 & 3 \end{bmatrix} \begin{Bmatrix} 1 \\ 1 \\ 1 \end{Bmatrix} = 3\begin{Bmatrix} 1 \\ 1.67 \\ 2 \end{Bmatrix}$$

For the second iteration, substitute $\begin{Bmatrix} \phi_{11} \\ \phi_{21} \\ \phi_{31} \end{Bmatrix} = \begin{Bmatrix} 1 \\ 1.67 \\ 2 \end{Bmatrix}$ into Eq. (6.67):

$$\frac{m}{k}\begin{bmatrix} 1 & 1 & 1 \\ 1 & 2 & 2 \\ 1 & 2 & 3 \end{bmatrix} \begin{Bmatrix} 1 \\ 1.67 \\ 2 \end{Bmatrix} = \begin{Bmatrix} 4.67 \\ 8.33 \\ 10.33 \end{Bmatrix}$$

After normalization, the above expression becomes:

$$\frac{m}{k}\begin{bmatrix} 1 & 1 & 1 \\ 1 & 2 & 2 \\ 1 & 2 & 3 \end{bmatrix}\begin{Bmatrix} 1 \\ 1.67 \\ 2 \end{Bmatrix} = 4.67\begin{Bmatrix} 1 \\ 1.79 \\ 2.21 \end{Bmatrix}$$

For the third iteration, substitute $\begin{Bmatrix} \phi_{11} \\ \phi_{21} \\ \phi_{31} \end{Bmatrix} = \begin{Bmatrix} 1 \\ 1.79 \\ 2.21 \end{Bmatrix}$ into Eq. (6.67):

$$\frac{m}{k}\begin{bmatrix} 1 & 1 & 1 \\ 1 & 2 & 2 \\ 1 & 2 & 3 \end{bmatrix}\begin{Bmatrix} 1 \\ 1.79 \\ 2.21 \end{Bmatrix} = \begin{Bmatrix} 5 \\ 9 \\ 11.21 \end{Bmatrix}$$

After normalization, the above expression becomes:

$$\frac{m}{k}\begin{bmatrix} 1 & 1 & 1 \\ 1 & 2 & 2 \\ 1 & 2 & 3 \end{bmatrix}\begin{Bmatrix} 1 \\ 1.79 \\ 2.21 \end{Bmatrix} = 5\begin{Bmatrix} 1 \\ 1.8 \\ 2.24 \end{Bmatrix}$$

For the fourth iteration, substitute $\begin{Bmatrix} \phi_{11} \\ \phi_{21} \\ \phi_{31} \end{Bmatrix} = \begin{Bmatrix} 1 \\ 1.8 \\ 2.24 \end{Bmatrix}$ into Eq. (6.67):

$$\frac{m}{k}\begin{bmatrix} 1 & 1 & 1 \\ 1 & 2 & 2 \\ 1 & 2 & 3 \end{bmatrix}\begin{Bmatrix} 1 \\ 1.8 \\ 2.24 \end{Bmatrix} = \begin{Bmatrix} 5.04 \\ 9.09 \\ 11.33 \end{Bmatrix}$$

After normalization, the above expression becomes:

$$\frac{m}{k}\begin{bmatrix} 1 & 1 & 1 \\ 1 & 2 & 2 \\ 1 & 2 & 3 \end{bmatrix}\begin{Bmatrix} 1 \\ 1.8 \\ 2.24 \end{Bmatrix} = 5.04\begin{Bmatrix} 1 \\ 1.8 \\ 2.25 \end{Bmatrix}$$

Convergence achieved and therefore, $\frac{1}{P^2} = 5.04\frac{m}{k}$. The frequency for third mode of vibration is $0.445\sqrt{\frac{k}{m}}$.

Figure 6.17 shows the mode shapes for first and third modes of vibration.

6.8   Rayleigh's Principle

By principle of conservation of energy, energy can neither be created nor destroyed and it can only be converted into different forms. When force is acting on a structure,

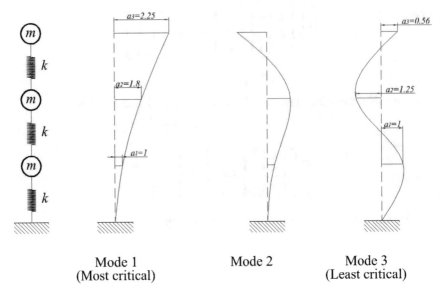

Mode 1
(Most critical)

Mode 2

Mode 3
(Least critical)

Fig. 6.17 Solution to Example 6.5

the structure will react accordingly in the form of displacement. Work is done and it is converted to energy. Without considering the damping of structure, while the structural elements behave within their elastic limit, energy will not be transferred outside the structure. Instead, it will remain in the structural system with the ability to change its form.

At any instant of time, the amount of energy in the system is constant. The types of energy that govern the behaviour of the structure, in this case, are kinetic and elastic strain energy. Kinetic energy, E_K is the function of mass and velocity, while elastic strain energy, E_{ES} is the function of stiffness and displacement.

$$E_K = \frac{1}{2}mv^2 \tag{6.68}$$

$$E_{ES} = \frac{1}{2}kx^2 \tag{6.69}$$

When the structure is vibrating, it possesses high kinetic energy and low elastic strain energy. When the structure has been displaced with its amplitude, kinetic energy inside the system is low and elastic strain energy is high. Therefore, the maximum amount of kinetic energy equals to that of elastic strain energy, while upholding the assumption where no energy loss from the system.

$$\frac{1}{2}mv^2 = \frac{1}{2}kx^2 \tag{6.70}$$

When a structure is in harmonic motion,

$$y = Y \sin Pt$$
$$\dot{y} = YP \cos Pt$$

P is the fundamental frequency of structure and Y is the amplitude of displacement. Under the most critical condition, $y = Y$ and $\dot{y} = YP$. By substituting these into Eq. (6.70) yields:

$$\frac{1}{2}m(YP)^2 = \frac{1}{2}kY^2$$

Simplify the equation above yields:

$$m(YP)^2 = kY^2$$

By eliminating the common term Y^2 yields:

$$mP^2 = k$$

By expressing fundamental frequency, P in terms of mass, m and stiffness, k yields:

$$P^2 = \frac{k}{m}$$

One of the main parameters to be determined during dynamic analysis of structure is the fundamental frequency. For multi-degree of freedom system, the fundamental frequency usually associated with the most critical first mode of vibration.

Consider a multi-degree of freedom system as shown in Fig. 6.18.

Under this condition, Rayleigh assumed each mass vibrates harmonically. In this case, the function of displacement and velocity are extended to take the response of all masses into consideration:

$$y_i = Y_i \sin Pt$$
$$\dot{y}_i = Y_i P \cos Pt$$

Under the most critical condition, $y_i = Y_i$ and $\dot{y}_i = Y_i P$. The kinetic and elastic strain energy in each mass can be determined by substituting the individual displacement and velocity into Eqs. (6.68) and (6.69), respectively:

$$E_{K,i} = \frac{1}{2}m_i \dot{y}_i^2$$
$$E_{ES,i} = \frac{1}{2}k_i y_i^2$$

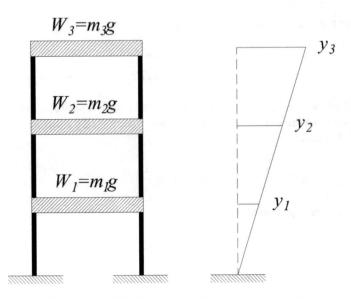

Fig. 6.18 Undamped two-degree of freedom system subjected to harmonic vibration

Kinetic and elastic strain energy of the system is the summation of individual mass' energy. Therefore,

$$E_K = \frac{1}{2} \sum m_i \dot{y}_i^2 \tag{6.71}$$

$$E_{ES} = \frac{1}{2} \sum k_i y_i^2 \tag{6.72}$$

The expression of kinetic energy can be rewritten by substituting $\dot{y}_i = Y_i P$ into Eq. (6.71) yields:

$$E_K = \frac{1}{2} \sum m_i (Y_i P)^2$$

By substituting $m_i = \frac{W_i}{g}$ into equation above yields:

$$E_K = \frac{1}{2} \sum \frac{W_i}{g} P^2 Y_i^2 \tag{6.73}$$

The expression of elastic strain energy can be rewritten by substituting $y_i = Y_i$ into Eq. (6.72) yields:

$$E_{ES} = \frac{1}{2} \sum k_i Y_i^2 \tag{6.74}$$

When force exerted to a mass, inertial force will be developed immediately and acts as reaction to the exerted force. Therefore, when an individual mass is displaced with amplitude Y_i

$$F = kY_i$$

Inertial force of an individual mass is defined as:

$$F = W_i = m_i g$$

By substituting $W_i = kY_i$ into Eq. (6.74) yields:

$$E_{ES} = \frac{1}{2} \sum W_i Y_i$$

By equalizing the maximum amount of kinetic energy with that of elastic strain energy yields:

$$\frac{1}{2} \sum W_i Y_i = \frac{1}{2} \sum \frac{W_i}{g} P^2 Y_i^2$$

Simplify the equation above yields:

$$\sum W_i Y_i = \sum \frac{W_i}{g} P^2 Y_i^2$$

By expressing P in terms of other variables yields:

$$P^2 = g \times \frac{\sum W_i Y_i}{\sum W_i Y_i^2} \tag{6.75}$$

Example 6.6 Application of Rayleigh's Principle

Solve for the natural frequency for structure in Fig. 6.15 using Rayleigh Principle.

Solution

From Example 6.4 Application of iterative method, the mass and stiffness matrices of the structure are as follow:

$$[m] = \begin{bmatrix} m & 0 & 0 \\ 0 & m & 0 \\ 0 & 0 & m \end{bmatrix}$$

$$[k] = \begin{bmatrix} 5k & -2k & 0 \\ -2k & 4k & -2k \\ 0 & -2k & 2k \end{bmatrix}$$

To use Rayleigh Principle, the displacement at each storey is crucial. In this case, they are required to be determined before performing the calculation.

$$[F] = [k][\Delta]$$

Since Rayleigh's Principle consider no energy loss within a system, the force developed in stiffness component equals to the force developed in mass component. Thus,

$$[W] = [k][\Delta]$$
$$[\Delta] = [k]^{-1}[W] \tag{6.76}$$

From Example 6.4 Application of iterative method, $[k]^{-1} = [f]$ and therefore:

$$[k]^{-1} = [f] = \frac{1}{6k}\begin{bmatrix} 2\ 2\ 2 \\ 2\ 5\ 5 \\ 2\ 5\ 8 \end{bmatrix}$$

Also, $[W]$ is the inertial force, which is a product of mass and gravitational acceleration. Therefore, Eq. (6.76) can be written as:

$$\begin{Bmatrix} \delta_1 \\ \delta_2 \\ \delta_3 \end{Bmatrix} = \frac{1}{6k}\begin{bmatrix} 2\ 2\ 2 \\ 2\ 5\ 5 \\ 2\ 5\ 8 \end{bmatrix}\begin{bmatrix} mg \\ mg \\ mg \end{bmatrix} = \frac{mg}{6k}\begin{bmatrix} 2\ 2\ 2 \\ 2\ 5\ 5 \\ 2\ 5\ 8 \end{bmatrix}\begin{bmatrix} 1 \\ 1 \\ 1 \end{bmatrix}$$

From matrix above,

$$\delta_1 = \frac{mg}{6k}(2 + 2 + 2) = \frac{mg}{k}$$
$$\delta_2 = \frac{mg}{6k}(2 + 5 + 5) = 2\frac{mg}{k}$$
$$\delta_3 = \frac{mg}{6k}(2 + 5 + 8) = 2.5\frac{mg}{k}$$

From Eq. (6.75):

$$P^2 = g \times \frac{\sum W_i Y_i}{\sum W_i Y_i^2}$$

By substituting the resultant displacement components into $\sum W_i Y_i^2$ and $\sum W_i Y_i$ yields:

$$\sum W_i Y_i = W_1 Y_1 + W_2 Y_2 + W_3 Y_3 = mg\left(\frac{mg}{k}\right) + mg\left(2\frac{mg}{k}\right) + mg\left(2.5\frac{mg}{k}\right)$$

$$= 5.5 \frac{m^2 g^2}{k}$$

$$\sum W_i Y_i^2 = W_1 Y_1^2 + W_2 Y_2^2 + W_3 Y_3^2$$

$$= mg \left(\frac{mg}{k}\right)^2 + mg \left(2\frac{mg}{k}\right)^2 + mg \left(2.5\frac{mg}{k}\right)^2 = 11.25 \frac{m^3 g^3}{k^2}$$

Therefore, the value for P^2 can be solved using the following equation:

$$P^2 = g \times \frac{5.5 \frac{m^2 g^2}{k}}{11.25 \frac{m^3 g^3}{k^2}} = 0.488 \frac{k}{m}$$

The natural frequency of structure is $0.699 \sqrt{\frac{k}{m}}$.

Example 6.7 Application of Rayleigh's Principle

Determine the base shear of structure under first mode of vibration as shown in Fig. 6.19 by referring to El Centro earthquake 1940. Take $\xi = 5\%$.

Solution

The mass and stiffness matrices for this structure are:

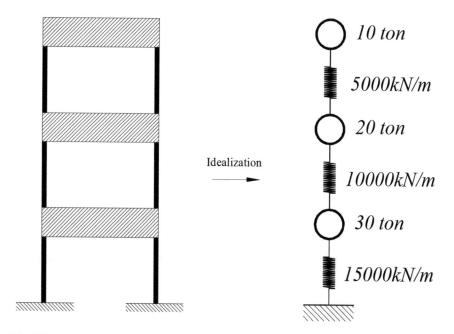

Fig. 6.19 Example 6.7

$$[m] = \begin{bmatrix} 30000 & 0 & 0 \\ 0 & 20000 & 0 \\ 0 & 0 & 10000 \end{bmatrix} = 10000 \begin{bmatrix} 3 & 0 & 0 \\ 0 & 2 & 0 \\ 0 & 0 & 1 \end{bmatrix}$$

$$[k] = 1000 \times 10^3 \begin{bmatrix} 15+10 & -10 & 0 \\ -10 & 10+5 & -5 \\ 0 & -5 & 5 \end{bmatrix} = 1000 \times 10^3 \begin{bmatrix} 25 & -10 & 0 \\ -10 & 15 & -5 \\ 0 & -5 & 5 \end{bmatrix}$$

From Eq. (6.76),

$$[\Delta] = [k]^{-1}[W]$$

Therefore, flexibility matrix is required. First, the determinant of $[k]$ is calculated as follows:

$$|k| = (1000 \times 10^3)^3 \{(25)[(15)(5) - (-5)(-5)] - (-10)[(-10)(5)]\}$$
$$= 1 \times 10^{18}[25(50) + 10(-50)] = 7.5 \times 10^{20}$$

Since the matrix is symmetrical, $[k] = [k]^T$. Next, the adjoint for $[k]$ will be determined as follows:

$$\text{adj}(k) = (1000 \times 10^3)^2 \begin{bmatrix} (15)(5)-(-5)(-5) & -(-10)(5) & (-10)(-5) \\ -(-10)(5) & (25)(5) & -(25)(-5) \\ (-10)(-5) & -(25)(-5) & (25)(15)-(-10)(-10) \end{bmatrix}$$

$$= 1 \times 10^{12} \begin{bmatrix} 50 & 50 & 50 \\ 50 & 125 & 125 \\ 50 & 125 & 275 \end{bmatrix}$$

The flexibility matrix is defined as follows:

$$[f] = [k]^{-1} = \frac{1}{|k|}\text{adj}(k) = \frac{1 \times 10^{12}}{7.5 \times 10^{20}} \begin{bmatrix} 50 & 50 & 50 \\ 50 & 125 & 125 \\ 50 & 125 & 275 \end{bmatrix}$$

$$= \frac{2}{15 \times 10^8} \begin{bmatrix} 50 & 50 & 50 \\ 50 & 125 & 125 \\ 50 & 125 & 275 \end{bmatrix}$$

By substituting the resultant flexibility matrix and relationship $W = mg$ into Eq. (6.76) yields:

$$\begin{Bmatrix} \delta_1 \\ \delta_2 \\ \delta_3 \end{Bmatrix} = \frac{2}{15 \times 10^8} \begin{bmatrix} 50 & 50 & 50 \\ 50 & 125 & 125 \\ 50 & 125 & 275 \end{bmatrix} \begin{bmatrix} 30000g \\ 20000g \\ 10000g \end{bmatrix}$$

$$= \frac{2g}{15 \times 10^8} \begin{bmatrix} 50 & 50 & 50 \\ 50 & 125 & 125 \\ 50 & 125 & 275 \end{bmatrix} \begin{bmatrix} 30000 \\ 20000 \\ 10000 \end{bmatrix}$$

From matrix above,

$$\delta_1 = \frac{2g}{15 \times 10^8}((50)(30000) + (50)(20000) + (50)(10000)) = 0.004g$$

$$\delta_2 = \frac{2g}{15 \times 10^8}((50)(30000) + (125)(20000) + (125)(10000)) = 0.007g$$

$$\delta_3 = \frac{2g}{15 \times 10^8}((50)(30000) + (125)(20000) + (275)(10000)) = 0.009g \quad (6.77)$$

From Eq. (6.75):

$$P^2 = g \times \frac{\sum W_i Y_i}{\sum W_i Y_i^2}$$

By substituting the resultant displacement components into $\sum W_i Y_i^2$ and $\sum W_i Y_i$ yields:

$$\sum W_i Y_i = W_1 Y_1 + W_2 Y_2 + W_3 Y_3$$
$$= 30000g(0.004g) + 20000g(0.007g) + 10000g(0.009g)$$
$$= 350g^2$$
$$\sum W_i Y_i^2 = W_1 Y_1^2 + W_2 Y_2^2 + W_3 Y_3^2$$
$$= 30000g(0.004g)^2 + 20000g(0.007g)^2 + 10000g(0.009g)^2$$
$$= 2.27g^3$$

Therefore, the value for P^2 can be solved using the following equation:

$$P^2 = g \times \frac{350g^2}{2.27g^3} = 154.2$$

Rayleigh's Principle always produces a fundamental frequency. Therefore, the fundamental frequency of structure is 12.42 rad/s and it is enough to proceed with next step.

The period of structure can be calculated using Eq. (3.15):

$$T = \frac{2\pi}{\omega_n} = \frac{2\pi}{12.42} = 0.51\,\text{s}$$

The displacement as shown in Eq. (6.77) is in fact for the first mode of vibration:

$$\delta_{11} = 0.004g = 0.039m$$
$$\delta_{21} = 0.007g = 0.069m$$
$$\delta_{31} = 0.009g = 0.088m$$

The resultant displacement matrix is:

$$\left\{ \begin{array}{c} \delta_{11} \\ \delta_{21} \\ \delta_{31} \end{array} \right\} = \left\{ \begin{array}{c} 0.039 \\ 0.069 \\ 0.088 \end{array} \right\}$$

The above results are normalized using Eq. (6.16):

$$\phi_{11} = \frac{\delta_{11}}{\sqrt{m_1\delta_{11}^2 + m_2\delta_{21}^2 + m_3\delta_{31}^2}}$$

$$= \frac{0.039}{\sqrt{30000 \times 0.039^2 + 20000 \times 0.069^2 + 10000 \times 0.088^2}} = \frac{0.039}{14.77}$$

$$= 2.64 \times 10^{-3}$$

$$\phi_{21} = \frac{\delta_{21}}{\sqrt{m_1\delta_{11}^2 + m_2\delta_{21}^2 + m_3\delta_{31}^2}}$$

$$= \frac{0.069}{\sqrt{30000 \times 0.039^2 + 20000 \times 0.069^2 + 10000 \times 0.088^2}} = \frac{0.069}{14.77}$$

$$= 4.67 \times 10^{-3}$$

$$\phi_{31} = \frac{\delta_{31}}{\sqrt{m_1\delta_{11}^2 + m_2\delta_{21}^2 + m_3\delta_{31}^2}}$$

$$= \frac{0.088}{\sqrt{30000 \times 0.039^2 + 20000 \times 0.069^2 + 10000 \times 0.088^2}} = \frac{0.039}{14.77}$$

$$= 5.96 \times 10^{-3}$$

$$\left\{ \begin{array}{c} \phi_{11} \\ \phi_{21} \\ \phi_{31} \end{array} \right\} = \left\{ \begin{array}{c} 2.64 \times 10^{-3} \\ 4.67 \times 10^{-3} \\ 5.96 \times 10^{-3} \end{array} \right\}$$

Based on $T = 0.51$ s and $\xi = 5\%$, by referring to Fig. 5.20 the following design values obtained for first mode of vibration:

$$S_a = 1g$$
$$S_d = \frac{S_a}{P^2} = \frac{1 \times 9.81}{154.2} = 0.064 \text{ m}$$

The mass contribution for the first mode of vibration is:

$$C_1 = \frac{m_1\phi_{11} + m_2\phi_{21} + m_3\phi_{31}}{m_1\phi_{11}^2 + m_2\phi_{21}^2 + m_3\phi_{31}^2}$$

$$= \frac{30000 \times 2.64 \times 10^{-3} + 20000 \times 4.67 \times 10^{-3} + 10000 \times 5.96 \times 10^{-3}}{30000 \times \left(2.64 \times 10^{-3}\right)^2 + 20000 \times \left(4.67 \times 10^{-3}\right)^2 + 10000 \times \left(5.96 \times 10^{-3}\right)^2}$$

$$= \frac{232.3}{1}$$

$$= 232.2$$

By using $z = [\phi][C_r] \times S_{dr}$:

$$z_{11} = 2.64 \times 10^{-3} \times 232.2 \times 0.064 = 0.039 \, \text{m}$$
$$z_{21} = 4.67 \times 10^{-3} \times 232.2 \times 0.064 = 0.069 \, \text{m}$$
$$z_{21} = 5.96 \times 10^{-3} \times 232.2 \times 0.064 = 0.089 \, \text{m}$$

By using $Q_{ir} = m_i\omega_i^2 z_{ir}$:

$$Q_{11} = 30000 \times 154.2 \times 0.039 = 180,414 \, \text{N}$$
$$Q_{21} = 20000 \times 154.2 \times 0.069 = 212,796 \, \text{N}$$
$$Q_{31} = 10000 \times 154.2 \times 0.089 = 137,238 \, \text{N}$$

The base shear can be determined with $\sum_{i=1}^{n} Q_{ir}$

$$V_1 = Q_{11} + Q_{21} + Q_{31} = 180,414 + 212,796 + 137,238 = 530,448 \, \text{N} = 530.4 \, \text{kN}$$

6.9 Holzer's Method

Holzer's method is commonly used in building with n-storey (n-degree of freedom). Holzer's method assumes the building behave like single degree of freedom system. By cutting the stiffness component at any storey, the lateral elastic restoring force developed in it equals to the summation of inertial force developed in the storeys above it. Consider a n-storey structure as shown in Fig. 6.20.

Say, the column located just below m_{i-1} is cut. Based on the consideration of Holzer's method, the shear force developed in it is:

$$F_{v,i-1} = \sum_{j=1}^{i-1} m_j\ddot{x}_j$$

At the same level, the elastic restoring force induced due to the relative displacement at lower storey.

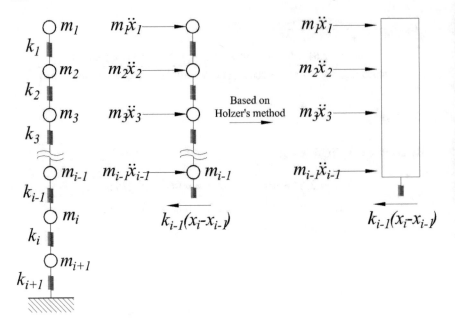

Fig. 6.20 Multi-storey building idealized for Holzer's method

$$F_{s,i-1} = (k_{i-1})(x_i - x_{i-1})$$

Under equilibrium, the shear force equals to the elastic restoring force:

$$\sum_{j=1}^{i-1} m_j \ddot{x}_j = k_{i-1}(x_i - x_{i-1}) \tag{6.78}$$

Assuming the vibration is harmonic, the structural response can be expressed as the following, where P denotes the natural frequency:

$$x = A \sin Pt$$
$$\dot{x} = AP \cos Pt$$
$$\ddot{x} = -AP^2 \sin Pt$$

By substituting the above expression into the Eq. (6.78) yields:

$$\sum_{j=1}^{i-1} m_j \left(-A_j P^2 \sin pt\right) = (k_{i-1})(A_i \sin Pt - A_{i-1} \sin Pt)$$

Rearranging the equation above and eliminate the common term $\sin Pt$ on both sides yield:

$$\sum_{j=1}^{i-1} m_j\left(-A_j P^2\right) = (k_{i-1})(A_i - A_{i-1})$$

$$\sum_{j=1}^{i-1} m_j\left(-A_j P^2\right) + k_{i-1}A_{i-1} = k_{i-1}A_i$$

$$\frac{\sum_{j=1}^{i-1} m_j\left(-A_j P^2\right) + k_{i-1}A_{i-1}}{k_{i-1}} = A_i$$

Rearrange the equation above yields:

$$A_i = A_{i-1} - \frac{p^2}{k_{i-1}} \sum_{j=1}^{i-1} m_j A_j \qquad (6.79)$$

Holzer's method works with the consideration where no displacement will occur at the base storey. The value of P^2 will be assumed and the analysis continues until the displacement at the base is obtained. If the residual, i.e. calculated base displacement is not zero or its deviation from zero is not acceptable, the trial and error process will need to be repeated. The displacement of the top storey is always taken as 1, since such displacement is normalized (Fig. 6.21).

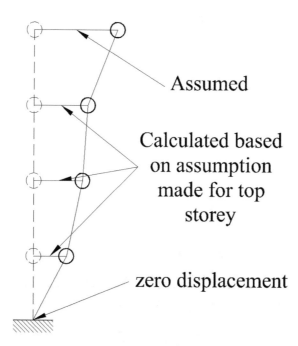

Fig. 6.21 Consideration made by Holzer's method in generalized mode of vibration

Table 6.1 Holzer's method table after step 1

Storey No.	A_i	m	mA	$\sum mA$	$\frac{1}{k_{i-1}}$	$\frac{p^2}{k_{i-1}}$	$\frac{p^2}{k_{i-1}}\sum mA$	$A_{i-1} - \frac{p^2}{k_{i-1}}\sum mA$
3		m_1			$\frac{1}{k_3}$			
2		m_2			$\frac{1}{k_2}$			
1		m_3			$\frac{1}{k_1}$			

To implement Holzer's method, the table is recommended to ease the proceed (see Table 6.1). Consider a structure as shown in Fig. 6.14, the Holzer's method can be carried out with the following steps:

(1) Determine the number of degree of freedom, and corresponding mass and stiffness components. The parameters for the topmost storey shall be filled in the top row of table, while the first storey shall be filled in the last row.
(2) Assume the natural frequency, P for the first trial.
(3) Calculate $\frac{p^2}{k}$ for each storey (Table 6.2).
(4) Assume the displacement at top storey. To ease the calculation, assume it to be 1 (Table 6.3).
(5) Calculate mA for current storey (Table 6.4).
(6) Find the sum of mA at current storey and all storeys above it (Table 6.5).
(7) Calculate the product of $\frac{p^2}{k}$ and $\sum mA$ for current storey (Table 6.6).
(8) Calculate the resultant displacement at lower storey by using Eq. (6.79) (Table 6.7).
(9) Bring over the calculated resultant displacement to the storey below and repeat Step 5 to 9 (Table 6.8).

Table 6.2 Holzer's method table after step 3

Storey No.	A_i	m	mA	$\sum mA$	$\frac{1}{k_{i-1}}$	$\frac{p^2}{k_{i-1}}$	$\frac{p^2}{k_{i-1}}\sum mA$	$A_{i-1} - \frac{p^2}{k_{i-1}}\sum mA$
3		m_1			$\frac{1}{k_3}$	$\frac{p^2}{k_3}$		
2		m_2			$\frac{1}{k_2}$	$\frac{p^2}{k_2}$		
1		m_3			$\frac{1}{k_1}$	$\frac{p^2}{k_1}$		

Table 6.3 Holzer's method table after step 4

Storey No.	A_i	m	mA	$\sum mA$	$\frac{1}{k_{i-1}}$	$\frac{p^2}{k_{i-1}}$	$\frac{p^2}{k_{i-1}}\sum mA$	$A_{i-1} - \frac{p^2}{k_{i-1}}\sum mA$
3	1	m_1			$\frac{1}{k_3}$	$\frac{p^2}{k_3}$		
2		m_2			$\frac{1}{k_2}$	$\frac{p^2}{k_2}$		
1		m_3			$\frac{1}{k_1}$	$\frac{p^2}{k_1}$		

Table 6.4 Holzer's method table after step 5

Storey No.	A_i	m	mA	$\sum mA$	$\frac{1}{k_{i-1}}$	$\frac{p^2}{k_{i-1}}$	$\frac{p^2}{k_{i-1}}\sum mA$	$A_{i-1} - \frac{p^2}{k_{i-1}}\sum mA$
3	1	m_1	m_1		$\frac{1}{k_3}$	$\frac{p^2}{k_3}$		
2		m_2			$\frac{1}{k_2}$	$\frac{p^2}{k_2}$		
1		m_3			$\frac{1}{k_1}$	$\frac{p^2}{k_1}$		

Table 6.5 Holzer's method table after step 6

Storey No.	A_i	m	mA	$\sum mA$	$\frac{1}{k_{i-1}}$	$\frac{p^2}{k_{i-1}}$	$\frac{p^2}{k_{i-1}}\sum mA$	$A_{i-1} - \frac{p^2}{k_{i-1}}\sum mA$
3	1	m_1	m_1	m_1	$\frac{1}{k_3}$	$\frac{p^2}{k_3}$		
2		m_2			$\frac{1}{k_2}$	$\frac{p^2}{k_2}$		
1		m_3			$\frac{1}{k_1}$	$\frac{p^2}{k_1}$		

Table 6.6 Holzer's method table after step 7

Storey No.	A_i	m	mA	$\sum mA$	$\frac{1}{k_{i-1}}$	$\frac{p^2}{k_{i-1}}$	$\frac{p^2}{k_{i-1}}\sum mA$	$A_{i-1} - \frac{p^2}{k_{i-1}}\sum mA$
3	1	m_1	m_1	m_1	$\frac{1}{k_3}$	$\frac{p^2}{k_3}$	$\frac{p^2 m_1}{k_3}$	
2		m_2			$\frac{1}{k_2}$	$\frac{p^2}{k_2}$		
1		m_3			$\frac{1}{k_1}$	$\frac{p^2}{k_1}$		

Table 6.7 Holzer's method table after step 8

Storey No.	A_i	m	mA	$\sum mA$	$\frac{1}{k_{i-1}}$	$\frac{p^2}{k_{i-1}}$	$\frac{p^2}{k_{i-1}}\sum mA$	$A_{i-1} - \frac{p^2}{k_{i-1}}\sum mA$
3	1	m_1	m_1	m_1	$\frac{1}{k_3}$	$\frac{p^2}{k_3}$	$\frac{p^2 m_1}{k_3}$	$A_2 = 1 - \frac{p^2 m_1}{k_3}$
2		m_2			$\frac{1}{k_2}$	$\frac{p^2}{k_2}$		
1		m_3			$\frac{1}{k_1}$	$\frac{p^2}{k_1}$		

Table 6.8 Holzer's method table after step 9 solved for second storey

Storey No.	A_i	m	mA	$\sum mA$	$\frac{1}{k_{i-1}}$	$\frac{p^2}{k_{i-1}}$	$\frac{p^2}{k_{i-1}}\sum mA$	$A_{i-1} - \frac{p^2}{k_{i-1}}\sum mA$
3	1	m_1	m_1	m_1	$\frac{1}{k_3}$	$\frac{p^2}{k_3}$	$\frac{p^2 m_1}{k_3}$	$A_2 = 1 - \frac{p^2 m_1}{k_3}$
2	A_2	m_2	$A_2 m_2$	$A_2 m_2 + m_1$	$\frac{1}{k_2}$	$\frac{p^2}{k_2}$	$\frac{p^2(A_2 m_2 + m_1)}{k_2}$	$A_3 = A_2 - \frac{p^2(A_2 m_2 + m_1)}{k_2}$
1		m_3			$\frac{1}{k_1}$	$\frac{p^2}{k_1}$		

(10) After the steps are performed for the first storey, check whether the resultant displacement at base, i.e. residual is acceptable. If not, repeat Step 2 with different natural frequency.

Example 6.8 Application of Holzer's method

Solve for the fundamental frequency for structure in Fig. 6.15 using Holzer's method.

Solution

From Example 6.4 Application of iterative method, the mass and stiffness for each degree of freedom are as follows:

$$m_1 = m, m_2 = m, m_3 = m$$
$$k_1 = 3k, k_2 = 2k, k_3 = 2k$$

Therefore,

$$\frac{1}{k_1} = \frac{1}{3k}, \frac{1}{k_2} = \frac{1}{2k}, \frac{1}{k_3} = \frac{1}{2k}$$

For the first trial, assume $P^2 = 0.3\frac{k}{m}$ (Table 6.9)

$$\frac{P^2}{k_1} = \frac{0.3\frac{k}{m}}{3k} = \frac{0.1}{m}$$

$$\frac{P^2}{k_2} = \frac{0.3\frac{k}{m}}{2k} = \frac{0.15}{m}$$

$$\frac{P^2}{k_3} = \frac{0.3\frac{k}{m}}{2k} = \frac{0.15}{m}$$

For the second trial, assume $P^2 = 0.4\frac{k}{m}$ (Table 6.10)

$$\frac{P^2}{k_1} = \frac{0.4\frac{k}{m}}{3k} = \frac{0.133}{m}$$

Table 6.9 First trial for Example 6.6

Storey No.	A_i	m	mA	$\sum mA$	$\frac{1}{k_{i-1}}$	$\frac{P^2}{k_{i-1}}$	$\frac{P^2}{k_{i-1}}\sum mA$	$A_{i-1} - \frac{P^2}{k_{i-1}}\sum mA$
3	1	m	m	m	$\frac{1}{2k}$	$\frac{0.15}{m}$	0.15	$1 - 0.15 = 0.85$
2	0.85	m	$0.85\,m$	1.85	$\frac{1}{2k}$	$\frac{0.15}{m}$	0.278	$0.85 - 0.278 = 0.572$
1	0.572	m	$0.572\,m$	$2.42\,m$	$\frac{1}{3k}$	$\frac{0.1}{m}$	0.242	$0.572 - 0.242 = 0.33$

Table 6.10 Second trial for Example 6.6

Storey No.	A_i	m	mA	$\sum mA$	$\frac{1}{k_{i-1}}$	$\frac{P^2}{k_{i-1}}$	$\frac{P^2}{k_{i-1}}\sum mA$	$A_{i-1} - \frac{P^2}{k_{i-1}}\sum mA$
3	1	m	m	m	$\frac{1}{2k}$	$\frac{0.2}{m}$	0.2	$1 - 0.2 = 0.8$
2	0.8	m	$0.8\,m$	1.8	$\frac{1}{2k}$	$\frac{0.2}{m}$	0.36	$0.8 - 0.36 = 0.44$
1	0.44	m	$0.44\,m$	$2.24\,m$	$\frac{1}{3k}$	$\frac{0.133}{m}$	0.298	$0.44 - 0.298 = 0.142$

$$\frac{P^2}{k_2} = \frac{0.4\frac{k}{m}}{2k} = \frac{0.2}{m}$$

$$\frac{P^2}{k_3} = \frac{0.4\frac{k}{m}}{2k} = \frac{0.2}{m}$$

For the third trial, assume $P^2 = 0.45\frac{k}{m}$ (Table 6.11)

$$\frac{P^2}{k_1} = \frac{0.45\frac{k}{m}}{3k} = \frac{0.15}{m}$$

$$\frac{P^2}{k_2} = \frac{0.45\frac{k}{m}}{2k} = \frac{0.225}{m}$$

$$\frac{P^2}{k_3} = \frac{0.45\frac{k}{m}}{2k} = \frac{0.225}{m}$$

For the fourth trial, assume $P^2 = 0.48\frac{k}{m}$ (Table 6.12)

$$\frac{P^2}{k_1} = \frac{0.48\frac{k}{m}}{3k} = \frac{0.16}{m}$$

$$\frac{P^2}{k_2} = \frac{0.48\frac{k}{m}}{2k} = \frac{0.24}{m}$$

Table 6.11 Third trial for Example 6.6

Storey No.	A_i	m	mA	$\sum mA$	$\frac{1}{k_{i-1}}$	$\frac{P^2}{k_{i-1}}$	$\frac{P^2}{k_{i-1}}\sum mA$	$A_{i-1} - \frac{P^2}{k_{i-1}}\sum mA$
3	1	m	m	m	$\frac{1}{2k}$	$\frac{0.225}{m}$	0.225	$1 - 0.225 = 0.775$
2	0.775	m	$0.775\,m$	$1.775\,m$	$\frac{1}{2k}$	$\frac{0.225}{m}$	0.399	$0.775 - 0.399 = 0.376$
1	0.376	m	$0.376\,m$	$2.151\,m$	$\frac{1}{3k}$	$\frac{0.15}{m}$	0.323	$0.376 - 0.323 = 0.053$

Table 6.12 Fourth trial for Example 6.6

Storey No.	A_i	m	mA	$\sum mA$	$\frac{1}{k_{i-1}}$	$\frac{p^2}{k_{i-1}}$	$\frac{p^2}{k_{i-1}}\sum mA$	$A_{i-1} - \frac{p^2}{k_{i-1}}\sum mA$
3	1	m	m	m	$\frac{1}{2k}$	$\frac{0.24}{m}$	0.24	$1 - 0.24 = 0.76$
2	0.76	m	$0.76\,m$	$1.76\,m$	$\frac{1}{2k}$	$\frac{0.24}{m}$	0.422	$0.76 - 0.422 = 0.338$
1	0.338	m	$0.338\,m$	$2.098\,m$	$\frac{1}{3k}$	$\frac{0.16}{m}$	0.336	$0.338 - 0.336 = 0.002$

$$\frac{P^2}{k_3} = \frac{0.48\frac{k}{m}}{2k} = \frac{0.24}{m}$$

The residual is acceptable after the fourth trial, when $P^2 = 0.48\frac{k}{m}$ is applied. Therefore, the fundamental frequency of the structure is $0.693\sqrt{\frac{k}{m}}$.

Example 6.9 Normal modes determination using Holzer's method

Determine the mode shapes for structure in Fig. 6.16 using Holzer's method. From Example 6.5 Normal modes determination using iterative method, it is known that natural frequencies for first and third mode of vibration are $0.445\sqrt{\frac{k}{m}}$ and $1.806\sqrt{\frac{k}{m}}$ respectively. Compare the result with the one obtained from Example 6.5 Normal modes determination using iterative method.

Solution

From Fig. 6.16, the mass and stiffness for each degree of freedom are as follows:

$$m_1 = m, m_2 = m, m_3 = m$$
$$k_1 = k, k_2 = k, k_3 = k$$

Therefore,

$$\frac{1}{k_1} = \frac{1}{k}, \frac{1}{k_2} = \frac{1}{k}, \frac{1}{k_3} = \frac{1}{k}$$

For the first mode of vibration, given $P = 0.445\sqrt{\frac{k}{m}}$ (Table 6.13)

$$P^2 = 0.2\frac{k}{m}$$
$$\frac{P^2}{k_1} = \frac{0.2\frac{k}{m}}{k} = \frac{0.2}{m}$$
$$\frac{P^2}{k_2} = \frac{0.2\frac{k}{m}}{k} = \frac{0.2}{m}$$

Table 6.13 Determination of mode shape for first mode of vibration using Holzer's method

Storey No.	A_i	m	mA	$\sum mA$	$\frac{1}{k_{i-1}}$	$\frac{P^2}{k_{i-1}}$	$\frac{P^2}{k_{i-1}}\sum mA$	$A_{i-1} - \frac{P^2}{k_{i-1}}\sum mA$
3	1	m	m	m	$\frac{1}{k}$	$\frac{0.2}{m}$	0.2	$1 - 0.2 = 0.8$
2	0.8	m	$0.8\,m$	$1.8\,m$	$\frac{1}{k}$	$\frac{0.2}{m}$	0.36	$0.8 - 0.36 = 0.44$
1	0.44	m	$0.44\,m$	$2.24\,m$	$\frac{1}{k}$	$\frac{0.2}{m}$	0.448	$0.44 - 0.448 = -0.008$

$$\frac{P^2}{k_3} = \frac{0.2\frac{k}{m}}{k} = \frac{0.2}{m}$$

For the first mode of vibration, given $P = 1.806\sqrt{\frac{k}{m}}$ (Table 6.14)

$$P^2 = 3.26\frac{k}{m}$$
$$\frac{P^2}{k_1} = \frac{3.26\frac{k}{m}}{k} = \frac{3.26}{m}$$
$$\frac{P^2}{k_2} = \frac{3.26\frac{k}{m}}{k} = \frac{3.26}{m}$$
$$\frac{P^2}{k_3} = \frac{3.26\frac{k}{m}}{k} = \frac{3.26}{m}$$

The normal modes are the same as the one obtained from Example 6.5 while having different values. This is because, in Example 6.5, the modal displacements are normalized with respect to the first storey, while in this example the displacements are normalized with respect to the top storey. By normalizing the displacements with respect to the first storey (Fig. 6.22):

Table 6.14 Determination of mode shape for third mode of vibration using Holzer's method

Storey No.	A_i	m	mA	$\sum mA$	$\frac{1}{k_{i-1}}$	$\frac{P^2}{k_{i-1}}$	$\frac{P^2}{k_{i-1}}\sum mA$	$A_{i-1} - \frac{P^2}{k_{i-1}}\sum mA$
3	1	m	m	m	$\frac{1}{k}$	$\frac{3.26}{m}$	3.26	$1 - 3.26 = -2.26$
2	-2.26	m	$-2.26\,m$	$-1.26\,m$	$\frac{1}{k}$	$\frac{3.26}{m}$	-4.11	$-2.26 - (-4.11) = 1.85$
1	1.85	m	$1.85\,m$	$0.59\,m$	$\frac{1}{k}$	$\frac{3.26}{m}$	1.92	$1.85 - 1.92 = -0.07$

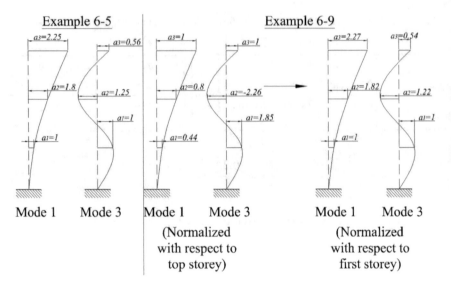

Fig. 6.22 Comparison of result obtained from Examples 6.5 and 6.9

6.10 Exercises

Exercise 6.1

Find the natural frequency and normal modes for the multi-degree of freedom system as shown in Fig. 6.23, considering $k = 0.25k_1 = 0.5k_2 = k_3$ and $m = 0.4m_1 = 0.4m_2 = m_3$.

Exercise 6.2

Determine the design shear force for the steel structure subjected to impact load as shown in Fig. 6.24. Ignore the effect of damping on the structural response and let the modulus of elasticity for steel be 200 GPa.

Exercise 6.3

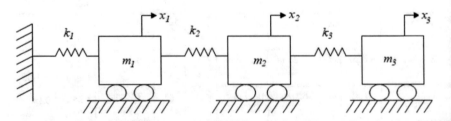

Fig. 6.23 Exercise 6.1

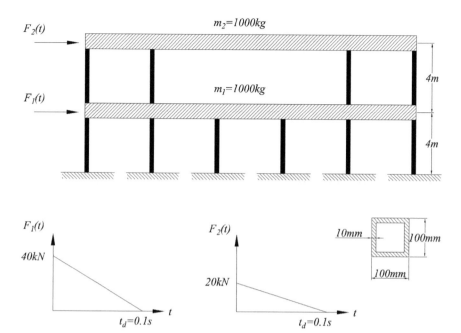

Fig. 6.24 Exercise 6.2

Find the base shear of two-storey of steel frame system shown in Fig. 6.25 when it's subjected to earthquake excitation noting that $\xi = 2\%$.

Exercise 6.4

Determine the natural frequency and mode shape for first and third mode of vibration for the structure as shown in Fig. 6.26 using Rayleigh's Principle.

Exercise 6.5

Determine the natural frequency and mode shape for first and third mode of vibration for the structure as shown in Fig. 6.26 using iterative method.

Exercise 6.6

Determine the natural frequency and mode shape for first and third mode of vibration for the structure as shown in Fig. 6.26 using Holzer's method.

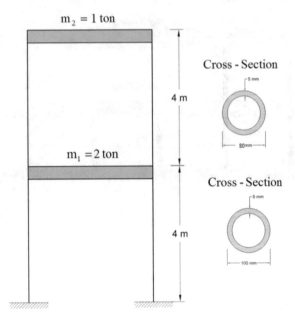

Fig. 6.25 Exercise 6.3

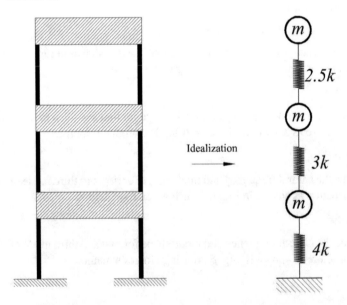

Fig. 6.26 Exercise 6.4, 6.5 and 6.6

Appendix

See Tables A.1 and A.2.

F. Hejazi and T. K. Chun, *Conceptual Theories in Structural Dynamics*,
Advanced Structured Materials 135,
https://doi.org/10.1007/978-981-15-5440-7

Table A.1 Geometric properties of various sections

Rectangular section	$A = bh$ $\bar{y} = \dfrac{h}{2}$ $I_{1-1} = \dfrac{bh^3}{12}$
Rectangular hollow section	$A = bh - (b - 2t_w)(h - 2t_f)$ $\bar{y} = \dfrac{h}{2}$ $I_{1-1} = \dfrac{bh^3}{12} - \dfrac{(b - 2t_w)(h - 2t_f)^3}{12}$
Triangular section	$A = \dfrac{bh}{2}$ $\bar{y} = \dfrac{h}{3}$ $I_{1-1} = \dfrac{bh^3}{36}$
Circular section	$A = \dfrac{\pi d^2}{4}$ $\bar{y} = \dfrac{d}{2}$ $I_{1-1} = \dfrac{\pi d^4}{64}$
Circular hollow section	$A = \dfrac{\pi d^2}{4} - \dfrac{\pi (d - 2t)^2}{4}$ $\bar{y} = \dfrac{d}{2} = r$ $I_{1-1} = \dfrac{\pi d^4}{64} - \dfrac{\pi (d - 2t)^4}{64}$

(continued)

Table A.1 (continued)

Semi-circular section	
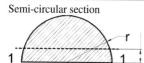 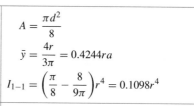	$A = \dfrac{\pi d^2}{8}$ $\bar{y} = \dfrac{4r}{3\pi} = 0.4244ra$ $I_{1-1} = \left(\dfrac{\pi}{8} - \dfrac{8}{9\pi}\right)r^4 = 0.1098r^4$
Parabolic section	
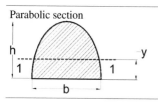	$A = \dfrac{2bh}{3}$ $\bar{y} = \dfrac{2h}{5}$ $I_{1-1} = \dfrac{8bh^3}{175}$

Table A.2 Structural member stiffness, k of member for various loading and restraint conditions

Simply supported beam with uniformly distributed load 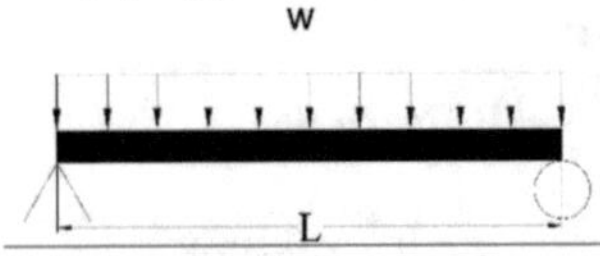	$k = \frac{384EI}{5L^4}$

w

L

| Simply supported beam with point load at midspan 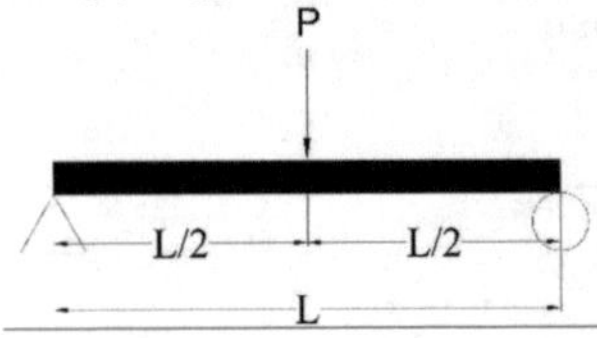 | $k = \frac{48EI}{L^3}$ |

P

L/2 ———— L/2

L

| Fixed ends beam with uniformly distributed load 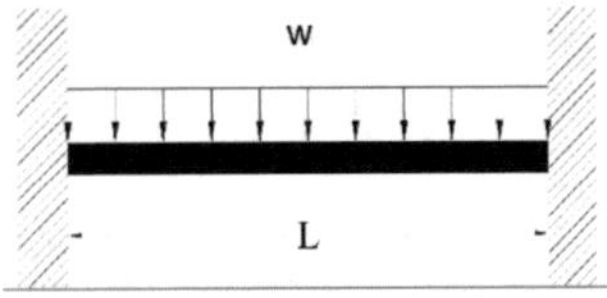 | $k = \frac{384EI}{L^4}$ |

w

L

| Fixed ends beam with point load at midspan 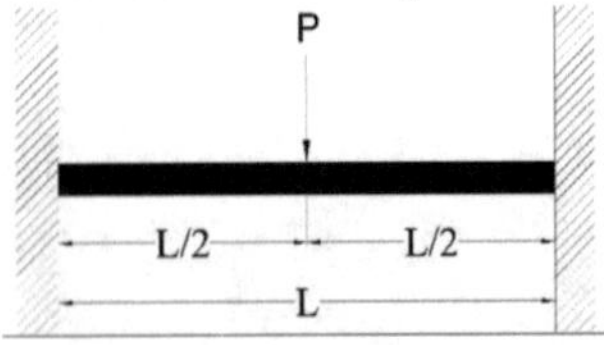 | $k = \frac{192EI}{L^3}$ |

P

L/2 ———— L/2

L

| Cantilever with uniformly distributed load 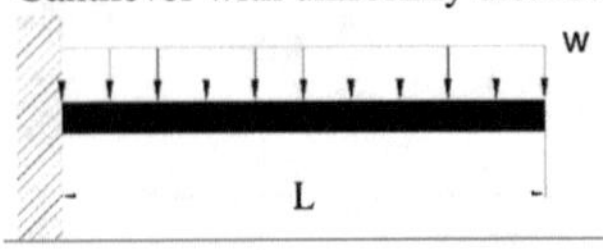 | $k = \frac{8EI}{L^4}$ |

w

L

| Cantilever with point load at free end 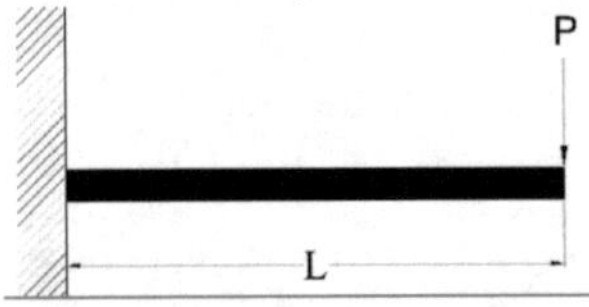 | $k = \frac{3EI}{L^3}$ |

P

L

| Fixed ends column 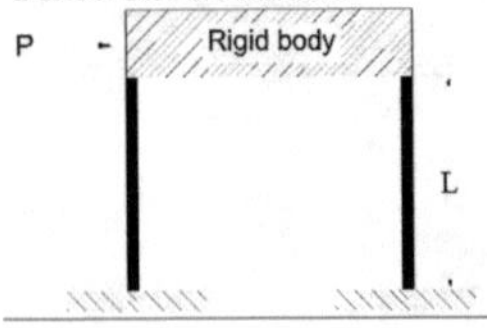 | $k = \frac{12EI}{L^3}$ for each column |

P — Rigid body

L

(continued)

Table A.2 (continued)

Pinned ends column	$k = \frac{3EI}{L^3}$ for each column

References

1. Hejazi, F.: Class Notes for "Structural Dynamics". Univerity Putra Malaysia, Malaysia (2019)
2. Chopra, A.K.: Dynamics of Structures Theory and Applications to Earthquake Engineering. Prentice-Hall, New Jersey, USA (1995)
3. Clough, R.W., Penzien, J.: Dynamics of Structures. McGraw—Hill, Inc., New York, USA (1993)
4. Paz, M.: Structural Dynamics: Theory and Computation. Van Nostrand Reinhold, New York, USA (1991)
5. Hart, G.C., Wong, K.K.F.: Structural Dynamics for Structural Engineers. John Wiley & Sons, New York, USA (2000)

The Editor(s) (if applicable) and The Author(s), under exclusive license
Springer Nature Singapore Pte Ltd. 2020
Hejazi and T. K. Chun, *Conceptual Theories in Structural Dynamics*,
Advanced Structured Materials 135,
https://doi.org/10.1007/978-981-15-5440-7

Printed in the United States
by Baker & Taylor Publisher Services